No. 942
$8.95

HULL CARE AND REPAIR

By
Dave MacLean

Blue Ridge Summit, Pa. 17214

FIRST EDITION

FIRST PRINTING—JUNE 1977

Copyright © 1977 by TAB BOOKS

Printed in the United States
of America

Reproduction or publication of the content in any manner, without express permission of the publisher, is prohibited. No liability is assumed with respect to the use of the information herein.

Library of Congress Cataloging in Publication Data

MacLean, Dave,   1921-
  Hull maintenance.

  Includes index.
  1. Boats and boating--Maintenance and repair.
2. Hulls (Naval architecture)--Maintenance and
repair. I. Title.
VM321.M25   623.84'028   77-3647
ISBN 0-8306-7942-1
ISBN 0-8306-6942-6 pbk.

TO RAYTHEON:
Where, but for their wisdom
I would be toiling yet
in anonymous mediocrity.

Preface

WHOM THIS BOOK WAS WRITTEN FOR AND WHY

If your boat is less than 30 feet long, chances are that this book is for you. If you are still smarting from a bill for minor repairs from the boat yard, then this book is for you. If you just bought what looked like an easy-to-read and understandable text on boat maintenance only to discover it was written in a language understandable only by a marine engineer-architect then this book is for you. If you love boats and the water on which they travel and you are determined not to be driven from them by those who seem bent on getting you out of boating, then this book is for you.

Often, because of the lack of extra money, I have been faced with either giving up boating or economizing by doing most of my own maintenance work. I pestered the experts and asked thousands of foolish (to them) questions...and I took notes. When I didn't understand their explanations which was often, I would try to translate them into simple uncluttered English and repeat them to the experts. Usually they would agree that my interpretation was correct, but just as often they would add, in a somewhat injured tone, "you make it sound so simple." Yet all I ever did was to filter out the salty lingo, the technical jargon, and the trade talk mumbo-jumbo that is used to keep the uninitiated from discovering how easy most maintenance jobs really are.

Over the years I have collected and recorded many trade and skill secrets which I'm going to pass on to you in this series of books. I want to help to keep you in boating because I am one of you!

In this volume and in the others to follow we are going to try to do something new. Essentially, this is a series of "how-to" books. They are much like many of the new automotive maintenance books that are filling book store shelves lately. These books were created to offset the poor quality and outrageous costs of car repair which has become a national concern. I had an extensive collection of Yacht (not boat) maintenance books and all of them seemed to have one common failing. Although each had been written by a recognized expert, nothing short of another expert (it seemed) could read and understand them. Some of them are downright insulting and at best patronizing. This series of books will not make that mistake if I can help it.

The maintenance tasks found in this book, whether they be preventive or corrective were first analyzed, usually by watching an expert do the job. Sometimes the actions were recorded on tape or film. Then they were studied or flow charted and finally written out in detailed steps. The written instructions, generally heavily illustrated were tried out using friends with lesser skills than the expert, to see if they could do the job. Errors and gaps in the steps were noted and the procedure was refined until anyone with only the most basic skills could do the job.

Naturally, there are some maintenance and repair jobs that you *should not* try to do yourself. Don't try to rewind the field coils of a starter motor or re-build an engine alone.

Yet there are many maintenance tasks that you *can* do yourself and in this series we'll discuss as many of these tasks as possible.

A BRIEF OVERVIEW OF THE CHAPTERS TO FOLLOW

In chapters I through IV we have included a lot of valuable and general information relative to the maintenance and repair of small craft. While you might be anxious to get into a chapter that applies to your boat, be sure to read the first four chapters. Then you need but to go to the chapter that applies to the type of hull material that your boat uses. For example, if you have a lap-straked wood hull, the chapter on the maintenance and repair on aluminum hulls can be of little interest to you.

In chapter I, we encourage you to look at your boat as a system, which, in turn, is made up of several subsystems. The whole concept of *planned preventive maintenance* is dependent on your adoption of this view of a boat. As a matter of fact, we have organized this series of books in such a way as to provide a

Fig. 1 None of these boats had a skipper with a Planned Preventive Maintenance Program!

how-to book for each of the subsystems that common sense tells us makes up a boat. In chapter I we detail each of the six subsystems and point out many of the maintenance soft-spots, as the maintenance engineering experts call them. Small craft hulls differ in design, construction, and the material. Because of this, both preventive and corrective maintenance differ from one to another.

Chapter II begins with a brief discussion of the advantages and disadvantages of the three major types of hulls. Since each type has its own maintenance and repair problems, it will help you to better understand the special needs of your boat's hull maintenance program.

Boats have many enemies and we'll describe each of them in detail and discuss cause and effect. This is followed by the detailed description of our proposed defense system, *Planned Preventive Maintenance* (PPM). When we have persuaded you to consider PPM as your primary defense against your boat's enemies we'll tell you how to plan a preventive maintenance program for *your* boat. However, we won't drop it there. We'll have some concrete suggestions for making a schedule that you can live with. We are going to do our very best to sell you on this idea because, if the Navy can make a billion-dollar atomic powered aircraft carrier work using this method, we can certainly make our little boats run better with the least cost and effort.

In chapter III the tools and materials required for a PPM program will be described. The tools will be kept to a minimum, for we have found (by watching the experts) that some tools can do many different jobs. A solid defense against or at least the control of rust, corrosion, and galvanic action (erroneously called electrolysis) will be described in this chapter, closing with the cause, effect, and control of both dry and wet rot.

In chapter IV we detail the tools, materials, and techniques for painting and varnishing, because a boat, much like a pretty girl, is always putting on or taking off paint.

In chapter V we discuss specifically what to do for a particular kind of hull material, beginning with the wood hull. Methods of planking a boat are described so you can better understand how to repair planks. Methods of fastening and an important discussion of the fasteners themselves follow, together with the kinds of wood used in the hull and elsewhere on the boat. Ribs (or frames if you are an expert) and stringers, deck supports, decks and miscellaneous parts will be described together with suggested methods of repair. Sections on seams and sealing methods, checking planking, and improving ventilation, fighting rot, checking hull fittings, and a section on bilges, brightwork, and deck maintenance complete chapter V.

Chapter VI departs from the preventive maintenance philosophy and covers specific *corrective* maintenance problems. We begin with general cutting and fitting techniques, then the repair of a rib section, and repair of a plank section. These particular repair techniques were selected because they are applicable to many other maintenance situations in which similar methods and procedures could be used, rather than trying to cover each and every possible repair that might have to be made on a wood hull.

Chapter VII combines both PPM and corrective maintenance for glass reinforced plastic (fiberglass) hulls and gives further details on the construction of GRP boats. Some problems caused by poor layup and construction are described with their cures and the maintenance of the gel coat or final finish of the boat. This is followed by some hints on working with fiberglass and plastic resin compounds.

In chapter VIII we will list and describe the advantages of metal hulls, principally those constructed of aluminum. Specific remedies for the reduction or control of galvanic action will be found in this chapter along with several corrective maintenance procedures for the cure of simple dents, severe dents, cracks and tears, and torn or loose rivets.

Contents

Chapter 1	YOUR BOAT IS A SYSTEM	1
Chapter 2	THE COMMON SENSE BEHIND PLANNED PREVENTIVE MAINTENANCE PROGRAMS	23
Chapter 3	TOOLS, MATERIALS, AND SOME TECHNIQUES AND METHODS FOR PPM	44
Chapter 4	PAINTING: THE FIRST LINE OF DEFENSE	63
Chapter 5	PLANNED PREVENTIVE MAINTENANCE AND THE WOODEN HULL	83
Chapter 6	CORRECTIVE MAINTENANCE AND THE WOOD HULL—REPAIRS REQUIRED DUE TO NEGLECT AND DAMAGE	94
Chapter 7	PLANNED PREVENTIVE AND CORRECTIVE MAINTENANCE AND THE FIBERGLASS HULL	122
Chapter 8	PLANNED PREVENTIVE MAINTENANCE AND CORRECTIVE MAINTENANCE FOR ALUMINUM HULLS	142
	INDEX	149

1
YOUR BOAT IS A SYSTEM

When facing a boat repair project, always keep in mind that the real reason most of us won't try something new is the fear of failure. Let me mention a few of the magnificent failures. How about Edison who, even after the thousandth try, still wasn't convinced that he *could not* invent a light bulb. Margaret Mitchell's novel, *Gone With the Wind,* had been turned down by every major publisher except the last one. Not long ago a man in California believed he could package and sell rocks as pets. Everybody laughed and called him a nut, they said, "You can't do that!" but he did and made a fortune. The only true failures are those who *never try at all!*

WHAT'S IN IT FOR YOU?

Consider the following. How would you like to be among the first with a boat back in the water at the beginning of the boating season. No, not someone who just pulls the cover off, launches the boat and takes off only to be towed in by the Coast Guard a few hours later. What I'm talking about is a skipper who launches his boat well ahead of everyone else and the boat is in tip-top, shiny bright condition. Adopting and carrying out a PPM program will help make you that skipper.

The average small craft owner, unlike the wealthy yachtsman, has limited time and money to spend getting his boat ready for the season, and then keeping it running during the season. A well planned program of *preventive maintenance* can very nearly stop all but the most unforeseen required *corrective* mainte-

2 Your boat is a system

nance. At that time of the year when you should be getting the boat ready for the season, a lot of other things are competing for your time and money.

A PPM program will conserve your limited time and funds and direct them toward producing the maximum effect in the shortest time with the least possible effort.

Next, let's face it, some of us aren't the kind of orderly thinkers and planners that we'd like to be around the boatyard. Don't confuse list making with organized behavior. If you think back through many of your boat maintenance projects—honestly—you'll surely recall a number of false starts and several unplanned trips for tools and materials. In addition, there were those unplanned and unwelcome interruptions due to weather, more urgent needs in other areas, and so on. For example, consider the skipper who thought he needed to replace the zinc anodes on his rudder and shaft and ended up having to replace both rudder and shaft because he failed to inspect and replace the zincs when he should have. Summing up then, a PPM program can reduce—if not eliminate altogether—much of time/money wasting running around and simple aggravation.

With a PPM program, potential problems are anticipated and *prevented before* they take place. This means that the boat is more often available for use—as a boat and not as a training ground for marine *corrective* maintenance. So PPM can increase boat availability time.

It almost goes without saying but we'll say it anyway: PPM makes the boat look better and perform better. The value of that alone will not be lost on anyone.

I know of few skippers who do not dream of trading up, whether to a newer boat, a bigger boat, or a better boat. PPM, faithfully practiced, cannot help but enhance the resale value of your boat. As a matter of fact it will be patently apparent to the potential buyer when he first views the boat and sees the results. Successful yacht brokers (boat salesmen) have told me that a boat is sold or not to a prospective buyer in the first *few seconds that he views the boat!*

While it may not be possible to put a dollar value on it, a boat maintained under a PPM program makes *you* look better. Among your peers, you acquire a reputation as a sharp boat owner. They begin coming to you for help with their boating problems.

And finally, a well maintained, safe, seagoing boat will help the skipper to achieve the title *all* skippers aspire to—Ancient Mariner!

YOUR BOAT'S SYSTEM

1. The Hull Subsystem

In Figure 1.1 we show a boat as a system made up of six subsystems. We will consider the hull subsystem first.

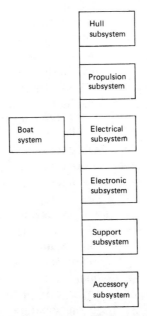

Fig. 1.1 The elements comprising a small craft system.

The boat's hull is shown in Figure 1.2. Naturally, individual boat hulls will vary in detail from one type of hull to another. But we can generalize here about where most of the troubles arise. These trouble spots are referred to as "soft spots." (This idea will be expanded in later chapters.) Assume we are looking at a wood-constructed boat for the moment. Starting at the bow, the *stem* serves as the sharp bladed point of the boat ("A" Figure 1.2). On a wood boat, this stem is made of the toughest, strongest wood that can be found, often white oak. The boat's planking attaches

4   Your boat is a system

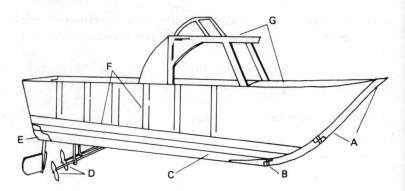

Fig. 1.2   Maintenance "soft spots" in a small craft.

to this one main hull member at the bow and often the method of planking will put great stress on the stem. It is first through the water as the boat goes ahead. Thus, anything in front of a moving boat—logs, buoys, docks, other craft—will hit the stem first. All floating obstacles seem to conspire to commit felonious assault on the stem.

There is yet another insidious enemy of stems and that is wood rot. Rarely do you ever *see* it until it is too late. We shall get somewhat tedious on the subject of wood rot in several later chapters of this book.

Still another of the spots on a boat that requires more attention than others is at that point where the stem is usually joined to the keel ("B" Figure 1.2). This is called the *scarph joint.* There is substantial stress on this joint, especially when the boat is underway. In part, the stress results from the normal working or movement of the boat's hull structure as it plows through the water. A boat is supposed to flex because, if it did not give slightly under stress of wind and wave it would very likely crack apart the way a rigid but fragile eggshell would.

Part of a planned preventive maintenance program will include inspection of the scarph joint each time the boat is taken out of the water or just before it is put back in.

Moving back toward the stern and bottom of the boat, the last planks of the skin that join to the keel or the keel's son (keelson) are called the "garboards" ("C" Figure 1.2). The garboards are very susceptible to damage and rot. In your PPM program you'll be keeping a sharp eye on these planks.

The stern area ("D" Figure 1.2) can be a seething trouble spot. If your boat is a regular inboard engine type, you can expect to spend a lot of time on propellers and shafts, shaft struts, zincs, stuffing boxes (where the propeller shaft comes through the hull), rudders, more zincs, rudder shaft boxes, and rudder posts. For boats with outdrives, the maintenance requirements alone would fill a book itself, and the same is true for boats with outboard engines. Small sailboats with inboard engines have the same maintenance requirements as inboard power boats.

Inside the boat and close to the waterline troubles often appear, mostly as a result of rot in wood boats in such areas as the braces for the transom called "knees" ("E" Figure 1.2). Rot will attack these as quickly as the garboards. Fresh water from rain and condensation is the carrier of the bacteria and fungus that cause wood rot. Neither of these demons lives well in salt water and fresh dry air.

Inside and out, rot can, and often does, attack a boat's planking and frames (ribs). The smallest opening of unprotected wood invites the heartbreak and shame of wood rot to make inroads. The engine bed (the wood stringers on which the engine is mounted) and floor timbers are very susceptible, for, like so many things below decks, they don't get their share of attention ("F" Figure 1.2).

Hard to believe yet true, the topsides around the cabin, windshield, hatches, and decks and deck combings are all attacked by rot, sunlight, — and, yes, man. Even though painted and otherwise protected from the drying of the sun, rain water seeps into the smallest nooks and crannies, under the paint, there to lie in wait, incubating a bad case of wood rot ("G" Figure 1.2).

Looking back to the hull subsystem and but a few of the soft spots in it can easily give the impression of a very black picture. Do not despair! You've got a lot going for you. Your PPM program will prevent or eliminate nine-tenths of these problems *before* they happen. Modern, durable and tough plastics, metals, protective compounds, and fasteners will nearly cover the rest. This will leave you with only *man* as the cause of unforeseen maintenance problems.

2. The Propulsion Subsystem

Figure 1.3 is a block diagram of a boat's propulsion subsystem. Naturally, it assumes a power-driven boat and not a sail boat. The block diagram, as shown, applies to virtually any engine including two- or four-cycle outboards, four-cycle inboards, or

6 Your boat is a system

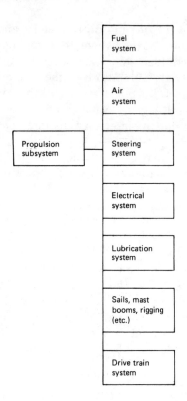

Fig. 1.3 The elements of a small craft propulsion subsystem.

two- or four-cycle diesels, and even one of the new gas turbines that are now available for marine use.

Beginning with the *fuel system,* we show in Figure 1.4 a sketch of a typical small craft fuel system. As a rule, fuel tanks cause few problems. Yet, as you can see, there are quite a few hoses and wires connected to the tank (or tanks). Normally, the hose connecting the fill fittings on deck to the tank is constructed of very strong flexible material. Because the hose is an electrical insulator, a grounding wire must be run from the tank to the metal deck fitting to guard against the accumulation of spark-generating static electricity. While there is little strain on the wire it is subjected to vibration and the terminals tend to corrode. The same trouble occurs with the electric wires that connect the fuel level pick-up inside the tank to the fuel gauge at the control station. The tank vents are normally connected from the tank top fitting by a hose to an overboard fitting well above

Your boat's system 7

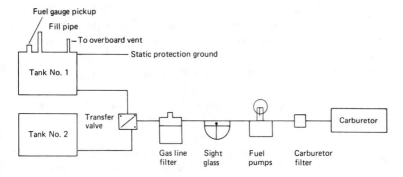

Fig. 1.4 "Typical" small craft fuel system.

the waterline. Vent lines have been known to clog with dirt and grease. When this happens the fuel pump on the engine will be struggling to pump fuel from the tank against a vacuum—something it cannot do. The engine dies, starved for fuel and, as a rule, it takes a long time to locate and cure this problem. In PPM, you will check and keep the vent lines clear so the problem will never occur.

Fuel tanks, whose manufacture and installation are, in part, regulated by the Coast Guard, do not often cause trouble. However, they may rust and corrode—then leak. Liquid gasoline is relatively harmless compared to gasoline as a vapor. It has been said that one cup of gasoline as a vapor has the explosive power of fifteen sticks of dynamite. Your PPM program will have this in mind.

A good boat has a primary fuel filter in the fuel line between the tank(s) and the fuel pump. Its job is to remove gross contaminants such as water, dirt and rust particles and pass only clear raw fuel. Fortunately, gasoline, unlike diesel oil, is not attacked by the strange bacteria which seems to thrive on diesel oil and causes the oil to clog filters, lines, and injectors. If you operate a diesel-powered boat in warm water, extra filters, tank clean-outs, and regular fuel treatments are a *must* in your PPM program.

The fuel pump, except for electrical pumps which are rare indeed on boats, is a diaphragm type of pump. Newer engines have a separate sight glass bowl, and, if any fuel appears in the bowl, it means that *one* of the *two* diaphragms in the pump has failed. While the pump will continue to operate, it must be repaired as *soon* as you get back to the dock. Older fuel pumps

8 Your boat is a system

do not have this safety feature and when they fail—that's it! If you are a PPM skipper you carry a spare. Some years ago, fuel pumps could be put back into good as new condition with a simple fuel pump rebuilding kit. This is no longer possible. Fortunately, fuel pumps, on the whole, are very reliable and give many hours of satisfactory performance.

Of the soft spots in a fuel system, the carburetor leads the list. In this day of high efficiency, low pollutant emission engines, it has become such a complex device that corrective maintenance is all but beyond the average skipper. Still, much *can* be done by the PPM skipper with a plan. He can clear and keep clear sticking chokes and fouled backfire flame arrestors. He can keep complex control linkage clean and lubricated. And he can make minor, but important, idle and idle jet adjustments. He can keep his fuel lines free of water and dirt and replace filter cartridges when required.

Figure 1.5 depicts the elements of an *air system* in a typical engine. The air system for any engine is normally free of repetitive problems. However, lack of a suitable mixture of fuel with air is very evident in the clouds of blue-black smoke from the engine exhaust. Only the supercharger found on diesel and hyped-up racing engines do not normally cause problems and will be of little interest or concern for small craft skippers. The backfire flame arrestor, which is mounted on the top of the

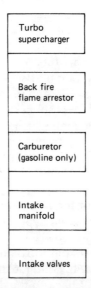

Fig. 1.5 Elements of a small craft engine air system.

carburetor's air intake throat, allows air to pass readily into the carburetor. Baffles are arranged in a manner to inhibit any flame coming *out* of the carburetor from a late firing or misfiring. This safety device is required by Coast Guard regulations on all *inboard* gasoline engines. In addition to the air which the engine draws in from the engine compartment, the oily vapors from the crankcase area and lube oil sump at the bottom of the engine are fed to the outside of the flame arrestor by hoses coming from the valve covers. The flame arrestor in normal operation quickly becomes coated with a layer of dirt and oil which tends to choke the air supply to the engine. The PPM skipper has a regular plan to prevent this.

The carburetor is part of both the air and the fuel systems and rarely causes trouble to the skipper who has a PPM and uses it. Intake manifolds are *almost* trouble free but they have been known to crack and leak.

Intake valves on the other hand *do* present occasional problems and little can be done of a preventive nature, other than to assure their proper adjustment and lubrication.

3. The Electrical Subsystem

The electrical system for small craft engines (Figure 1.6) is so loaded with soft spots and has so many natural enemies that we

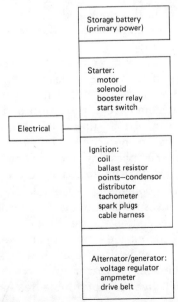

Fig. 1.6 Elements of a small craft electrical system (engine only).

10 Your boat is a system

have written a separate book to cover them all. Most small craft, even those propelled by outboard engines, are dependent upon an automative type lead-acid storage battery for all the electrical loads required. Because the battery is a chemical engine that converts chemical energy to electrical energy, it is subjected to many stresses. This is especially true when the boat in which it is installed is operated on salt water. Salt in any form is an anathema to a lead-acid battery. It is an understatement to say that the battery is crucial to the operation of a boat. Most inboard engines cannot be started without one, and many of the big powerful outboards can no longer be manually started with a pull cord except by King Kong. Yet the PPM skipper will have no trouble with his battery for he will be aware of all of the things that *might* cause trouble and has taken sensible precautions to see that they don't.

The *starter system* is equally susceptible to the slings and arrows of outrageous misfortune. Whenever a system uses and requires the huge amounts of electric current that the starter motor does—and at a very low voltage, we might add—there is bound to be trouble.

As for the *ignition system,* of all the soft spots in a boat this has got to be the softest. Ninety-nine out of a hundred problems around a boat seem related to the ignition system. There are just too many to detail in this book and we shall devote an entire chapter to ignition maintenance in the book on Small Craft Electrical Maintenance.

As for the alternator or generator we are in a bit of luck here. Ever since alternators were made available to boats as well as automobiles, many of the problems associated with the battery charging system have, for the most part, disappeared. It should be kept in mind that alternators satisfactory for automotive applications are not considered safe to use aboard a small craft. On the back of the marine alternator, the electrical connection that controls the output of the machine (the field winding connection) is completely covered. Thus, should there be any sparking at the field winding brushes, it would not cause the ignition and resulting explosion of any stray gasoline fumes in the engine compartment. Keep in mind that you will pay dearly for this little safety feature should you have to repair or replace your alternator. An alternator, properly provided with PPM, will give years of trouble-free service.

Another bit of good news for small craft owners is the appearance of transistorized, fully encased and potted voltage regulators. (After the unit has been assembled and placed in its case, a potting substance is poured in, which, when it hardens, completely surrounds, insulates, and fully protects the electrical components from every misadventure but one). Since there are no moving parts to wear out in a transistorized regulator and it is protected from everything but excessive heat, theoretically, it should last forever.

Small craft must have an indicator that informs the skipper of the current (no pun intended) state of the battery charging system. This can be an ineffectual device. It is far preferable to have an ammeter, or even better, a combination of two meters, a voltmeter *and* an ammeter. These devices seldom seem to cause much trouble and they do not require maintenance other than an occasional cleaning and tightening of the terminal lugs of the wiring connected to them.

We will not discuss the boat's electrical service system here other than to say it, like the ignition system, is loaded with soft spots and we shall devote a solid section of the Small Craft Electrical Maintenance Book to it.

4. The Lubrication Subsystem

The lubrication system (Figure 1.7) is not known to be a trouble maker on boats. In an inboard engine, as long as the oil level in

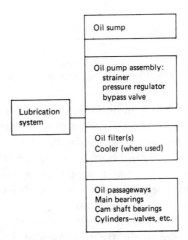

Fig. 1.7 Elements of a small craft lubrication system (engine only).

12   Your boat is a system

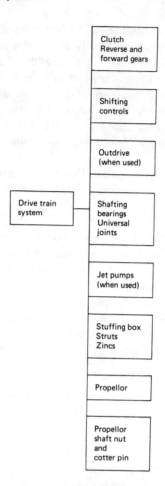

Fig. 1.8 Elements of a small craft drive train system.

the sump is periodically maintained at a proper level, the filter(s) and oil changed according to a plan, then the oil system will take care of itself. Not included in the diagram are the numerous points around a small craft that require but a drop of oil here, a dab of vasoline there—and a shot from a grease gun on that fitting once in a while—all according to a plan which *makes sure* each point needing it, gets lubrication attention.

In Figure 1.8 are shown the gross elements that make up the drive train for a typical small craft. Many of the problems found in this system are caused by the lack of a sensible and conscientiously practiced lubrication plan. While such a system

Your boat's system 13

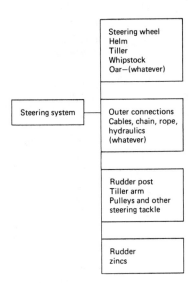

Fig. 1.9 Elements of the steering system.

is under great stress when pushing a boat through the water, rugged parts and good design (for the most part) will keep the system trouble-free when properly maintained with lubrication and adjustment according to a plan.

5. The Steering Subsystem

Figure 1.9 depicts most of the elements in a small craft steering system. Variations in design and style from boat to boat are often so different that no fixed system can be shown. In general, most small craft, whether inboard or outboard, will use a system of cables, pulleys, and springs or the more expensive so-called rack and pinion single cable system.

To say that the steering system, like the battery, is crucial is an understatement. A PPM inspection and maintenance program will keep the steering system steady.

6. The Life Support Subsystem

Figure 1.10 shows the life support system on a small craft. Perhaps "comfort and convenience system" might be a more proper term. Over the years most of the elements of this system have gone through a thorough design and tryout period. Those elements that have survived do their duty on small craft, as a rule, quite well. Some of the more recent innovations such as air

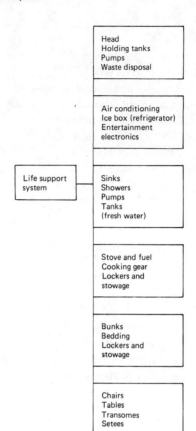

Fig. 1.10 Elements of a small craft life support system.

conditioners, electrical refrigerators, and microwave ovens are, for the most part, yet to be adapted to go to sea. As a result they have ended up creating a maintenance nightmare. A majority of them require normal 110 volts of house current which is still not readily available aboard most small craft. Regrettably, some skippers are rigging their boats for dock power to supply all the convenience equipment brought aboard. To us, this seems to tie the boat to the dock with a near unbreakable umbilical cord, for who wants to go fishing *without* the air conditioning on *all* day?

In any event, keep in mind that the more comfort and convenience equipment brought aboard, the greater your maintenance load. The more mundane elements, such as stoves, sinks, and heads, get the most use and will, as a consequence, require the most maintenance.

7. The Safety Subsystem

Figure 1.11 shows our view of the typical small craft safety system. Of course not all boats will have everything shown. However, a good safe boat will have a lot of them.

Most of the equipment shown in the diagram is rarely needed (hopefully)—for which we may be thankful. As a result, it just sits there and deteriorates unless a persistent and well planned PPM program keeps it ever ready for any emergency.

Soft spots are apt to occur in fire extinguishers, since they cannot be tested except when they are due for recharging. Now they are required to have gauges to indicate their state of charge and readiness.

The new smoke and fume detectors have built-in test features and rarely give trouble when adjusted and tested periodically according to a plan.

All pumps, such as those for fire, flooding, cooling, and water supply, require a lot of maintenance. You would be amazed at the number of pumps you have aboard if you were to add up every single one. They all require periodic attention.

Your greatest problems with flotation equipment maintenance comes (it seems) from small children whose attraction to emergency equipment seems irresistible. Keep this in mind when children are aboard.

Lights of any kind aboard a boat, whether they be installed, fixed, or portable, are a definite soft spot requiring maintenance. This is especially true if the boat is operated on salt water. As previously described, problems are caused by low voltage (battery operated), corroded connections, dampness, and vibration.

Most of the signaling equipment is primitive and not normally subject to failure or the need for much maintenance. The same is not true of any of the electronic equipment which often requires *extensive* corrective maintenance done only by expensive, licensed professionals. Piloting and navigation equipment is included for those skippers who like to cruise long distances, even in a small craft. There is very little the average small craft skipper can do in the way of maintenance on this kind of equipment except get it to the doctor when it is sick.

We include ground tackle (overall term for anything that makes up the anchoring gear) as it can and often does serve as safety equipment. It gets hard use and needs constant care. We would call it a soft spot if only to rivet your attention to it for its inclusion in your PPM program.

16 Your boat is a system

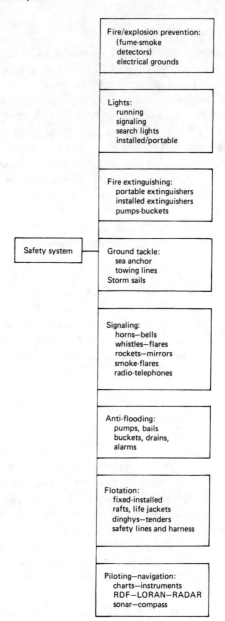

Fig. 1.11 Elements of a small craft safety system.

8. The Electrical Subsystem

The elements in a typical small craft electrical system (excepting engine electrics) are shown in Figure 1.12. Some boats

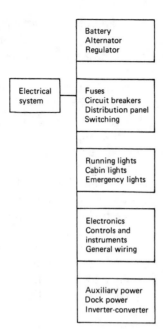

Fig. 1.12 Elements of a small craft electrical system.

may have much less, others may have more. Every year, as more and more designers and manufacturers enter the recreational boating field, the competition between them provides us with more and better equipment. Much of this equipment is electrical in nature or it is activated by electric power. Because of this the average skipper is forced to become generally knowledgeable of *practical* electrical principles and practices—unless he is well heeled enough to hire an electrical engineer to repair all his equipment each time he leaves the dock. Electrical equipment requires materials that are good conductors of current such as copper and aluminum. Both metals are said to be "active," which means that other metals and chemicals dearly love to work on them to form new, and hardly welcome, chemical substances. For example, that green patina that forms on the surface of unprotected copper terminals is formed by the combination of oxygen in the air and the copper. The result is a cuprous oxide that is very resistant to the flow of electric current. If there is also a salt content in the air, the chemistry becomes more complex and the picture for bad electrical connections darkens. Only a PPM program has any hope of winning out against this insidious enemy.

18 Your boat is a system

We have already spoken of the boat's battery as one of the principle soft spots in the engine or propulsion system. Suffice it to say that in small craft it has many duties and must receive tender loving care.

A heavy gauge wire usually leads from the battery to the small craft's control panel. At or near this panel the wire will connect to a fuse block. This is an insulated block on which there are mounted a number of fuse clips. This block provides a distribution and tie point for all the various electric loads throughout the boat. A better and more elaborate central distribution might use one or more of the panels shown in Figure 1.13. Normally these fuse blocks and distribution panels give reliable service without excessive maintenance requirements other than occasional fuse replacement and cleaning and tightening of terminal lug nuts. Much depends, of course, on the care and layout of the original installation. You must not expect your boat wiring and its accessibility to maintenance to come from the builder looking in any way like that of the *custom* professional job shown in Figure 1.13.

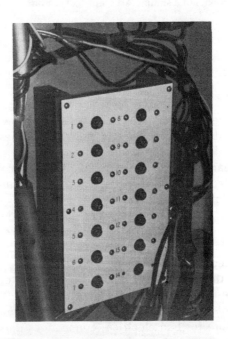

Fig. 1.13 Small craft electrical distribution panel with reset-type circuit breakers. Photo also shows proper wiring (note cable ties).

In addition to the fuse block or distribution panel, wise and cautious small craft skippers are going to sea these days with two batteries aboard instead of relying on one crucial battery. When properly installed and connected, they are not simply connected in parallel which for a number of reasons is bad news. Instead they are joined through a large control switch having several positions—OFF for completely disconnecting the batteries from all circuits, BOTH to connect both batteries in parallel, BATTERY #1 to connect one battery into the load, and BATTERY #2 to connect the second battery into the load. This installation is a great improvement over that of a single battery where, if anything should go wrong, you can't start the engine. However, the switch shown in Figure 1.14 is not inexpensive and must be installed with proper gauge wire and heavy duty terminal lugs. In addition, the use of this switch depends a lot on the skipper's memory of which battery he used the last time he went out so that each battery is kept fully charged. Also, the switch must be manually operated. A better idea comes from the electronic industry. This is the dual battery isolator (DBI) shown in Figure 1.15. This unit, when installed, is completely automatic,

Fig. 1.14 Example of a dual battery control switch (note positions).

Fig. 1.15 Example of a dual battery isolator that makes the system automatic.

has no moving parts, and permits the use of two batteries, each of which provides current to different loads, yet can be charged from *one* alternator at the same time. Thus, one battery could be reserved *solely* for the purpose of starting the engine and is fully charged at all times, while the other battery can serve the balance of the boat's electrical needs. DBI's normally sell for about half the cost of the more complex switches.

A small craft's lighting system (for the most part) is regulated by Federal law which in turn is based upon International law, at least as far as the boat's running lights are concerned. If the boat is to be operated at night or in low visibility situations, common sense and law dictate that certain lights be displayed when the boat is under way, and others when it is at anchor. In addition, there are the comfort and convenience lights such as cabin and cockpit illumination, safety and piloting lights (such as spot lights and bow eye lights, etc.). All of these require a program of PPM. Normally they are not much trouble if it is kept in mind that they are subjected to corrosion at all connections. In addition to these lights there are often such requirements as chart lights, instrument lights for keeping an eye on engine performance at night, compass light, and the like.

These days even the smallest of small craft are beginning to carry a significant load of electronics. Small depth sounders and fish finders, radiotelephones, radio direction finders, fume and smoke detectors, loud hailers, electronic ignition systems, loran receivers, and even radar can be found on the smallest boats. That these machines require maintenance, often extensive, goes without saying. The radiotelephones, by law, cannot be worked on except by qualified and licensed technicians. Most of the balance of the equipment requires extensive technical knowledge and elaborate instruments for all but the most rudimentary preventive maintenance. Yet, the skipper can do a lot to keep his electronic gear in good operating shape. Fortunately, space age and military electronics have provided a better than average degree of reliability for most of these units and a program of PPM designed to keep out salt atmosphere and compensate for excessive shock and vibration will do much to keep them out of the shore-side shops.

As more and more comfort and convenience electrical devices come aboard small craft, larger demands are made on the boat's primary power system. The ordinary battery simply cannot supply these demands. As a result we see more and more small craft with auxiliary power supply systems, such as engine-driven generators and inverters or converters. More and more boats are appearing with dock-power installations. Some skippers find to their dismay that these installations cause more problems than they solve. The dangers of galvanic action seem to increase tenfold when the boat is on dock power. The dangers from fatal electrical shock increase and, if that were not enough, the skipper can spend an inordinate amount of time maintaining all the extra equipment that is attracted aboard as soon as 110 volts of AC become available. It is safe to say that for most skippers, the disadvantages of auxiliary power often outweigh the advantages. The maintenance requirements and problems associated with small craft electrical systems necessitate a separate text which is available.

9. The Support System

Figure 1.16 depicts the oft forgotten (by many skippers) small craft's support system. If your boat is launched from a trailer and stored on the same trailer when not in the water, or if it is stored on a cradle during the off season, then these require maintenance. The trailer requires a lot of preventive maintenance.

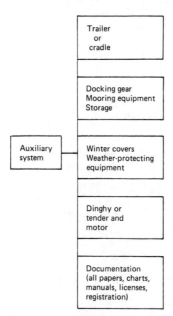

Fig. 1.16 Elements of a small craft support system.

If the boat is kept at a slip or dock or if it swings on a permanent mooring then there is more equipment to be maintained. Mooring buoys, anchors, dock lines, washdown hoses, gear lockers, bait boxes, and spare parts not carried on the boat all seem to require attention. How much and when is best controlled through a PPM plan.

In the off season in more northern climes, winter covers, dehumidifying systems, lashing-down lines, and framing for winter covers all demand time and attention. Dinghies, tenders, or inflatable rafts and the kicker motor that goes with them are on the list. Finally, one thing that rarely gets proper attention is the documentation requirements—these include such elements as registration and indentification documentation, radio licenses, engine and piloting logs, guest registers (for tax deductions), fuel purchases sales slips (for the same purpose), engine maintenance manuals, charts and other piloting publications, compass deviation tables, equipment maintenance manuals, and on and on. Yet you'll agree that these must be maintained and kept up to date.

2
THE COMMON SENSE BEHIND PLANNED PREVENTIVE MAINTENANCE PROGRAMS

If, in chapter I, we had some success in convincing you that a boat can be looked at as a system, which in turn is made up of several subsystems then the next thing to consider is the WHY of preventive maintenance. As we have pointed out, each of the systems may be thought of as having various soft spots which seem to be the trouble-makers for maintenance. The first impression marks a very dark one, but that's not truly the case. Consider for the moment the case of the atomic-powered submarines which perform their function so well. Because they are so well maintained, it takes *two* full crews to keep them functioning.

The *blue* crew takes them out to sea on patrols lasting as long as three months and returns to port. After supplies are loaded, the *gold* crew immediately takes the sub back to sea for another three months. Considering where and in what this "boat" must operate, how is it kept going so long as to need *two full crews*? The answer lies in a maintenance philosophy that has as its goal the *prevention* of material failure *before* it happens.

It is this requirement for maintenance that (along with other things) has driven a lot of skippers out of boating. Too many skippers, faced with a boat maintenance requirement of either the preventive or corrective nature, feel as handy as a paraplegic octopus. Beaten before they begin, they often do not begin at all.

In this chapter we shall discuss the three basic hull types, advantages and disadvantages and the related problems which are legion. We shall then describe an approach and a method of planning for Preventive Maintenance. We know of many skippers who are otherwise very competent boatkeepers but who seem to run from one brush fire to the next, never quite catching up with the work long enough to relax and just enjoy the boat for any length of time. Planning will stop that.

TYPES OF HULLS

There are three basic types of hulls and assuming you have but a passing interest in marine architecture we shall pass over these quickly. Understanding the type of hull you have and its advantages and disadvantages will have more bearing on your planning than the actual maintenance.

The Displacement Hull

The oldest and most common type of hull configuration is the displacement hull. The cave man with his hollow log, the Phoenecians, Romans, Vikings, and Columbus all had craft with displacement hulls. This type of hull floats low in the water and displaces a volume of water equal to its weight. It has advantages and applications, as you might expect. The displacement hull plows through the water and the design of the hull is such that it is said to have a certain hull speed, which means that no matter how much more power is added to the propulsion system the boat will not go faster than its hull speed. Increased engine power merely causes the stern to "squat" deeper in the water and burn up more fuel without any resultant increase in speed. However, when operated at their rated cruising speed this type of boat takes relatively small amounts of power which can be sustained in poor sea conditions and for long periods of time. This kind of hull would be preferred by the long-distance cruising skipper. It is more stable in rough water, but has a distinct tendency to roll (because of the rounded bottom). Figure 2.1 is an example of the displacement hull. You should not assume that

Fig. 2.1 Example of a small craft displacement type of hull.

because so much of the hull is under water that it will require more maintenance because that is not the case.

The Planing Hull

The second, and one of the most popular types of hulls, is the planing hull. This hull does not "swim" or "plow" through the water. Instead it is designed to "skim" or "plane" over the top of the water. Figure 2.2 illustrates a planing hull. At slow speeds this type of hull squats down by the stern, using considerable power and fuel, and leaving a tremendous wake. However, once enough power is applied to get it up on the "plane" (or "step"), it will proceed at substantial speeds. In a well designed boat, power can be actually reduced and the boat will stay on the plane very efficiently. The planing hull permits greater width or beam in the design and in turn allows the designer to create more useful space inside the boat. Although they have far less tendency to roll in a beam sea where the waves hit on the side of the boat, they are very rough riders in anything but a flat calm. In rough water they pound, are very noisy, and are not as stable as the displacement hull. They are most suitable on lakes, rivers, and short cruises in shallow water.

The "Vee" Hull

By far the most popular type of hull of the present day is the "Vee" hull available either as the modified Vee or the Deep Vee. These hulls were developed from the ocean power boat racing hulls. These high-speed boats were designed to bash through the

Fig. 2.2 Example of a small craft planing hull—near flat bottom.

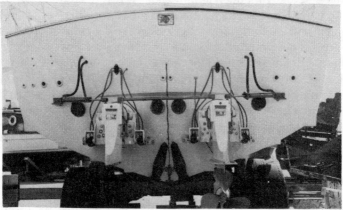

Fig. 2.3 Examples of a small craft deep vee planing hull.

roughest ocean waters. They could take such rough water that the drivers and mechanics had to be strapped into their seats. Out of this competition, as so often happens, came the Vee hull found in boats from sixteen feet to sixty feet in length. The Vee hull permits the design of fairly stable long distance cruisers which can still get up to a very respectable speed with good sea while keeping characteristics of greater stability.

In spite of what you might expect the maintenance requirements are about the same for each of these three kinds of hull. Even though the displacement hull appears to have a greater surface of the hull in the water all the time, both the deep vee and the flat bottomed planing hull still have a lot of surface area in the water except at high speeds. While the displacement hull cannot operate in as shallow water as the others, there is less danger (at least to the cautious skipper) of striking the propellers and shafts. Then too, the faster the boat the greater the damage to the hull when unseen obstructions such as sunken logs, snags and rocks are struck. The older displacement hulls often have a large skeg (a sort of keel) that protects the propeller, drive shaft, and rudder against underwater damage.

THE "ENEMIES" OF BOATS

Whether the boat planes or plows, it may seem to you that every living thing in salt, fresh, or brackish water is at war with your boat's hull. Let's take a careful look at the enemies one by one. If you know the enemies you can better construct a defense against them.

Sunlight

As you'll recall from your high school science classes, sunlight contains various levels of energy. At low energy levels are found the beneficial infra-red wave lengths. However, proceeding upwards in the sunlight spectrum we arrive, eventually, to the very high energy levels of ultra-violet light. While neither of these extremes are visible to the human eye, the results of their effects are very much in evidence. The destructive effects of ultra-violet (UV) light rays are evident on your skin after a long day in the boat or on the beach. This happens often even on a cloudy day when it is seemingly safe to lie around without protective clothing on. The energy content of these rays is so high that they penetrate the cloud layers and can deliver a painful sunburn. While you can get under cover, your boat can't—it has to sit there and take it. Everything topside on the boat and unprotected from the sun is bombarded with trillions of high energy UV rays. Some of the effects are the dulling of that glossy sheen of paint on a wood boat or the rapid oxidation of the gel coat on a fiberglass hull. UV rays working together with the oxygen in the air strike the atoms of the surface coat on the boat raising the energy levels of those atoms and causing them to form new chemical combinations (usually undesirable ones). They attack the pigments in both paint and gel coats. What they will do to the exposed varnished brightwork is a crime. The induced heat causes all materials to expand and as it cools in the evening they contract. This means that something is happening all the time. If not watched and corrected beforehand, small cracks in the paint form to admit fungus and bacteria.

Air

You might be surprised that we regard air an enemy of boats. Some aspects of air are of course beneficial. Noisy air that fills the sails and moves the boat is certainly desirable. The cool breeze that makes the hot day liveable and dry air that evaporates the dampness in the hull and storage lockers are welcome. But, what about the air that carries the spoors of wood rot fungus, or the bacterial form of wood rot? Air also carries the dust and dirt that spoil the appearance of the boat and must be washed off lest dirt become imbedded in the pores of the gel coat or paint. Drying air is the same air that carries the moisture, that, under the right temperature conditions, precipitates out on the surface of the boat as a dew, thus helping to begin the work of wood rot fungus. By far the most insidious aspect of air is free oxygen. Oxygen has an affinity for any other atom available. The resulting oxygen combinations are seldom welcome. The dull and faced gel coat on a fiberglass boat has been oxidized. Free oxygen also attracts trouble on metal surfaces. Oxygen prefers to bond with metallic atoms and iron is one of its favorites. Aboard a boat any metal having the slightest iron content is a target for oxygen. When the oxygen atom and the atom of iron combine the result is rust (technically known as ferric oxide). However, as bridge, ship, and engine builders have found, the near perfect protection against rust is a coat of RUST! It seems that rust forms so densely on the metal surface that little additional rusting action occurs. Summing up, air can be a mixed blessing, depending on how you look at what it is doing to your boat.

Salt

This ordinarily harmless substance usually means a lot of trouble for most skippers. Even those boats with skippers who are lucky (or smart) enough to keep them in fresh water are not totally immune from the ravaging effects of salt, because in fresh water too there is *some* salt, not in the concentration as in sea water and in a different chemical form (such as metallic salt), but salt nevertheless. When salt water evaporates on the various surfaces of a boat it leaves behind crystals of salt, which, working with the UV rays of the sun, will have a bleaching effect on everything. In contact with metal, salt forms metallic salt combinations that help accelerate rust and corrosion. Where copper is carrying electric current, salt will form high-resistance connections that are difficult to find and cure. Salt around an engine is just plain bad news. The engine compartment *must* be ventilated to let the engine breathe and to eliminate the accumulation of dangerous fumes. The normal amount of heat present dries the air and leaves the salt crystals behind to do their dirty work. The result of salt water accidentally spilled on or near a lead acid battery and into the battery to combine with the sulfuric acid is *chlorine gas!* If inhaled by the skipper or crew chlorine gas causes a painful and permanent shortness of breath. Yet it is just barely possible to find some good in salt. For example, the stringers and frames in the bottom of a wood boat that are kept wet with salt water will benefit from a pickling effect; wood rot will never occur, for the bacteria and fungus of rot cannot live in well soaked salted wood. Summing up, the only place for salt on a boat is on the skipper's hat and in his language. However, salt in solution with water becomes most destructive because then it is an electrolyte or a liquid capable of conducting an electric current which brings us to our next enemy.

Galvanic Action

Often erroneously called "electrolysis," galvanic action is the eating away of a metal fixture on a boat at a rate far more rapid than the ordinary corrosive actions of rust and salt. This happens when two different kinds of metals, close to each other, are both immersed in an electrolytic liquid. Salt water is an electrolyte as it is a good conductor of electric current. Over the years it has become common practice to call this electrolysis (although to say this is in error is splitting technical hairs). However, to be precise, electrolysis is the action that takes place within a lead-acid storage battery when under a charging current. The passage of the charging current through the battery's electrolyte (a mixture of sulphuric acid and water) causes some of the molecules of water to separate into their constituent atoms of oxygen and hydrogen. These form gas bubbles which float to the top of the battery to be emitted from the battery vents. The hydrogen gas is very explosive and must be carefully vented. All flame and sparks should be kept away from a battery that is on charge. Galvanic action is the fundamental principle behind any battery—or to be more precise, a primary cell. Two or more cells make up a battery. To oversimplify the cell principle, when two metals, not alike, are immersed in an electrolyte, a current will flow between the two metals. The direction of that current depends on which of the two metals is the more "noble," meaning less active. Silver is more noble than copper and copper is more noble than iron. One of the least noble of metals, practically a serf, is zinc. Zinc will readily give up not only current, but whole atoms and molecules of zinc to a more noble metal such as bronze, copper, aluminum, or steel. When this happens, the zinc is rapidly eaten away by *sacrificing* itself to a noble cause. As a matter of fact, zinc is used extensively as a "sacrificial" metal and is used as a replaceable anode to protect propellers, shafts, struts, outdrives, and oil coolers. Even iron exhaust manifolds can be protected with zinc anodes. Galvanic action together with wood rot will cause the average skipper the most concern and require the most maintenance, whether it be preventive or corrective.

WOOD ROT

Wood rot and galvanic action have one thing in common. Their attack on the boat and its fittings is, for the most part, insidious and often unnoticed until it is too late. Skippers must know these enemies and be on guard against them at all times. Just because you might have a fiberglass boat does not mean that you will not have to fight galvanic action or wood rot. Galvanic action is just as common a problem on fiberglass hulls as it is on any other hull—sometimes more so. As for wood rot, fiberglass boats still have a lot of wood fixtures, trim, and other wooden parts aboard, all of which are targets of the bacteria and mold spoors of wood rot.

Fresh water, either rain or dew or even the water that condenses on the inside of a fiberglass hull is the carrier of wood rot. The bacterial and fungus spoors (seeds) are carried by fresh water and spread in the air. A boat stored in a boat yard can "catch" rot from a pile of discarded lumber scraps that have been allowed to rot. The parent fungus that causes the worst and most rapid form of rot can spread or give off *one billion* spoors, which can start yet another case of rot, each by itself. Both forms of rot eat and digest the fiber of the wood leaving behind a punky, stinky, useless substance that will infect everything around it or near it. When it is at work and the air is still you can smell it. Its odor is the only thing we have going for us. The easily recognizable musty rotting odor says, "Get out the ice pick or hammer and start hunting the source." Once started wood rot spreads rapidly.

Just as cancer cells break off and start new cancers in the human body, so does wood rot dispatch its spoors out to start new rot spots. The cure for wood rot is surgery. First, locate the infected spot, determine its dimensions and cut out the infected area of wood. Next, a new piece of wood is cut, fitted, treated with rot protection chemicals, and installed. While bacteria and fungus do not live in salt water they will attack all natural fibers such as cloth canvas, bedding, clothing, foodstuffs, water, and even paint. For those fortunate skippers who have boats in southern waters year round, there is the ever present danger from a weird bacteria that thrives in diesel fuel. This bacteria actually grows and multiplies in fuel oil and can choke filters, fuel lines, and injectors if allowed to. Do not despair—just knowing about the enemies is half the battle.

PLANTS AND ANIMALS

Brace yourself for this strange fact: some 600 different plants and 1300 species of animals have been identified that attack boats! All these attach themselves to the bottom of a boat, there to grow and multiply until the boat can no longer move. Among these critters is the unlovable barnacle which attaches itself to the hull with a glue that is the envy of man. The barnacle can be scraped from the hull, but takes a piece of the hull with him when he goes. Fortunately for us the defense against plants and animals (not counting rot) is near perfect when planned and applied properly. The answer is anti-fouling paint and we shall devote a goodly amount of space to this later on.

MAN

You might find it odd that we would consider man as an enemy of boats. But, is not man the direct cause of boat damage to boats requiring corrective maintenance? Does he not often neglect his responsibilities to provide proper and timely preventive maintenance thus causing needless wear and tear on his boat? Man's stubborn stupidity, persisting in failing to learn anything about the proper and safe way to operate and keep a boat, is basically the cause of expensive and often needless corrective maintenance. Without belaboring the point, it would seem that man can indeed be considered an enemy of boats, second only to the barnacle.

DEFENSES

Now that you know more of the enemies of boats let us open the discussion on defense. In this book and the others to follow, we will describe in all the detail that space allows the various defensive measures you can take to preserve, protect, and repair your boat. Yet all of this will be worthless unless we can persuade you to at least give the PPM system a try. Even PPM may not work if you are *not sold on it yourself*. That's the reason behind all this hard sell. The plan MUST be *yours* and *yours* alone—it must be so unique to you and your boat that it is unlikely to work for another boatman and his boat. There is a good reason behind this. The PPM system had been developed and perfected experimentally in the Navy and as long as the designers were around, it worked like a charm! But as soon as they left, the troops went back to their old sloppy disorganized ways and ships soon became maintenance nightmares. A second long-term study revealed that the men who had to make the plan work had not been "sold" on it's value to them. To this day a concerted effort is being made to convince everyone involved in ship maintenance that it is *his* program. It works beyond the best expectations. There is no way that we can sell you on this program—you have to sell yourself. You may agree that small craft skippers have very limited resources that must be spread to the maximum effectiveness. This can only be done by formulating an effective plan that will require time, money, limited skills, and materials to maintain a boat in a safe and handsome operating condition. The PPM system goes like this. Plans that work are always based on "hard" data. Therefore, the first step in the PPM system is to gather all the factual data available on *your* boat's maintenance needs.

THE PPM PROGRAM

The PPM program is a three-step process. These steps are separate, but closely related. First, study your boat's maintenance requirements, then develop your own common sense maintenance plan—and, carry it out!

Begin by applying the motto, "Inspect, don't expect!" While you can use the Owner's Manual that comes with the boat, don't depend on it alone, for most manuals are limited to a periodic lubrication program and assume the rest of the boat will take care of itself.

The Hull

Start with an inspection of the hull of the boat. Divide the hull in two parts. There's enough to watch for both inside and outside to deserve your undivided attention. Since the outside seems to get most of the wear and can get you into trouble faster, it deserves immediate attention. The outside of the hull can be again divided into two parts at the water line—the part of the hull that is under water and the remainder of the hull. In order to make a thorough, critical inspection that will gather *all* the facts you'll need the following:

1. An inspection checklist
2. A flashlight for those dark corners
3. A sharp pointed tool, such as a knife, or ice pick
4. A small hammer for tapping of suspected spots
5. A set of screwdrivers of assorted sizes
6. Bright colored chalk for marking bad spots

Now that you've started listing things, take time to put together your inspection checklist. Start at the bow and work back toward the stern both inside and outside the hull. A well thought out list composed before you start will help to make sure that you check everything that may need attention later. If your boat is still in the water or if it is not convenient to look at the actual boat, use a photo of the boat as a guide and reminder. By studying the

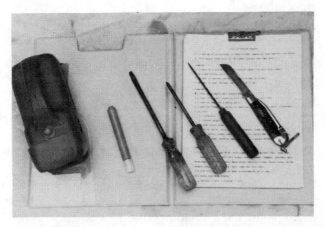

Fig. 2.4 Tools and supplies needed for preplanned maintenance.

34 The common sense behind planned preventive maintenance programs

Fig. 2.5 Stem area on a small craft displacement hull.

photo, you can mentally walk around the outside of the boat, reminding yourself and adding to the inspection list. For example:

THE STEM (if the boat is a wood hull)

1. How's the paint?
2. Any loose fasteners at the ends of strakes (planks) and the stem?
3. Any sign of rot where the strakes join the stem? (Test with sharp-pointed tool.) Check all suspicious openings between joints where fresh water could creep in and set.
4. If the stem has a metal guard on its leading edge, what shape is it in?
5. Any loose screws or other problems in the stem guard that need tightening or straightening?
6. In the scarph joint, where the keel joins the stem, are there any loose bolts, unusual cracks, or separations? Test with sharp point. Since this joint takes a lot of stress, check it out thoroughly.

Fig. 2.6 This illustration shows newly replaced plank in the garboards (lowest planks) on a small craft displacement hull.

Fig. 2.7 Carefully inspect condition of last year's bottom paint.

7. Now the garboards (the first planks of the skin and lowest down next to the keel). They are first to suffer when the boat is grounded accidentally or while beaching and ramping. In addition they are most susceptible to rot from the fresh water that collects in the bilges. It is often only possible to test from the outside of the boat due to obstructions inside the boat. Test the entire length of each garboard by tapping with a small hammer or testing with pointed tool.

8. Evaluate last season's anti-fouling paint, overall first. In general, how well did it do in keeping marine growth from attaching itself to the hull? You might be surprised to know that there is a wide choice in the selection and application of anti-fouling bottom paint these days, and that what works well in one area and on one kind of boat may not work as

well on another. We shall address this most important factor in the chapter on painting.

Now the bottom paint in detail. Any large patches missing? What about cracks and flaking? If the old coat looks good but the bottom is still badly fouled it may not be the fault of the paint at all, just the wrong paint for your boat and boating area.

9. Now check the condition of the keel, skeg, external ballast, centerboards, exposed bolt or screw heads. On the underside of the boat, carefully examine all of the metal fittings that stay underwater when the boat is afloat. Pay close attention to the condition of the sacrificial zinc anodes. They *should* be partially eaten away and, generally, must be replaced each season. If they are in good shape, they may *not* be doing their job and we need to find out why. Any sign of galvanic action on any exposed metal? Check stuffing box nuts for tightness because they can back off from vibration. Go over the whole of the rudder assembly with a fine tooth comb. A boat that will not steer is a distinct embarrassment to any skipper! If you have an outdrive unit, it is worth a checklist of its own. Be sure to go over all the lubrication points, looking for any evidence of damage, and be most critical of the propeller, its hub, lock nut, thrust washer—everything.

The preceeding are but suggestions of things that you should consider in making out a checklist. Again, we urge you not to

Fig. 2.8 This general stern area on a small craft displacement hull indicates that there is much work to be done.

make out the checklist by looking at your actual boat, for if you do, you will find that you will become diverted, discouraged, and demotivated. If you have done the checklist properly, you'll have covered the outside of the hull below the water line from stem to transom. Your own checklist will be different from this model. It will certainly be longer and in far more detail. A complete list for a normal size boat should take several sheets of paper. As you begin the actual checking, the list may grow longer as you uncover new points that you overlooked before. One of the reasons to list in advance is that it's easier to add items as you go along.

When preparing your inspection list be sure to leave ample space for notes and comments. One of the many good things about the composition of the check-off inspection list is that once it has been completed to your satisfaction, it can be used again and again. It can be handwritten in the sketchiest of notes. You must *not* get the impression that the composition of the checklist and the inspection is just more work to be added to your maintenance burden. If you'll bear with us a little longer we will show you that overall, this system is going to *save* a lot of wasted effort, time and money.

You will find as you use the checklist that it can be expanded and sometimes is out of order. Keep adding to it as needed and rearrange it as you go along. Be sure to save the check-off list because you're going to want it next year. It might be a good off-season project to incorporate the inspection check-off list in the boat's log book.

After composing inspection check points for the outside section of the hull above the water line, prepare a list for the inside of the hull, dividing it into convenient increments. Next, a separate check-off list should be composed for each of the systems suggested earlier in chapter I. The best time for all of this activity is after the boat has been hauled for the season and bedded down. This way you can still be involved in a crucial boating activity even though you can't be out on the boat. For those fortunate skippers who boat in warmer climes and only have to take the boat out of the water for annual overhaul, the inspection check-off list is an invaluable tool, especially if you are paying for the time the boat is on the ways at haul-out.

With the inspection check-off list completed and the inspection done, the hard data-gathering step is finished. Now is the time to *analyze* the collected data and put it in an order sensible to you. The order for doing the required preventive or corrective maintenance is clearly indicated, but what tools and materials are needed, how long should you allow for each task, and how should you schedule each task? Some further analysis will determine what work YOU can do and what must be allocated to the boatyard experts.

The next step in the PPM system is the actual planning phase. You have now determined what has to be done from the collected and ordered inspection data and you are now ready to actually *plan* each job. Determine (if possible) the procedure to follow and what materials and parts will be needed. These items should be estimated and carefully listed. *There is nothing so wasteful of time and energy as having to stop and procure forgotten materials, tools, and parts in the middle of a job!!!* Don't just plan, make contingency plans. What if it rains on the day you had scheduled for varnishing the cockpit brightwork? Have an alternative task that can be done, even if it is raining.

By analyzing what has to be done and how, you are better able to determine whether or not you can do it yourself or if it must be handled by an expert. It is this *job planning* that saves the most time, aggravation, and money. It will help get your boat in the water early in the season. Unless you have so much free time that you can do things when they need doing, you must sort your duties out in advance. By job planning you cut out most of the wasteful running about, getting ready to do the job.

With every job check-listed and analyzed, you are ready to assign priorities to everything. This part of planning counts heavily because some jobs just can't be completed until the previous job before them is done. Consider weather in your planning. If it is rainy or too humid to paint or varnish, then today's the day you can trace, mark, and repair those small leaks in the cabin roof, ports, or hatches.

You have the job data all collected and analyzed, and you know what your priorities are. Next, set up the *schedules*. This means allocating one of the most valuable of resources the average skipper has. We assume you are not a professional yachtsman and cannot devote full time to maintaining your boat. Thus, the time that *can* be spared must be conserved and utilized fully. Carefully estimate the time needed for each job. Set a specific and *realistic* goal to be accomplished in the time estimated. One of the most frustrating things for many boat owners is the discovery that they have set themselves almost impossible tasks, allowing half the time it would take to do the work correctly. The kind of planning we suggest will get it done, on time, with the least possible pain. It's no good to schedule a job for an afternoon when realistically it requires an entire day. Occasionally you'll finish one job in less time than anticipated and feel like you're really rolling. Stifle the urge to immediately tackle the next major task. My advice is: quit on a peak. Start the next job the next morning when you are rested, fresh, and well organized. Working sensibly and intelligently you'll be amazed at how fast all the maintenance jobs can be done and how soon you'll be in the water enjoying your well maintained boat.

The following is an example of a small but important section of a typical PPM plan and is offered here to demonstrate that the procedure is neither as complicated nor as difficult as you might have imagined. The example is *not* specifically for any *one* boat but most of the features will be found relevant to all boats. Not every single plan that you or I make in this process is going to work or suit you. Ruthlessly discard those that don't, but first find out *why* they didn't work. Maintenance engineers agree that *negative* information is valuable data and can be used to formulate *positive* plans. This is how you might generate a check inspection list, collect data, conduct a job analysis, list tools and material, and create a schedule for the application of anti-fouling paint just before launching.

Sample below-waterline checklist

1. What is the condition of last season's coat of anti-fouling paint?
2. If generally fair or poor, is the remaining paint:

    a. washed off in large patches? (indicating need for undercoat preparation)
    b. heavily coated with dried growths, slime, barnacles? (indicating need to consider different type and quality of paint)
    c. badly flaked? (indicating need to perfect application technique)
    d. chipped, with bare spots and bubbles? (indicates careless groundings and possible application of paint over wet or waxy surface)

    NOTE: With these questions answered, the blame can be placed on the paint, the painter, the hull, or conditions and an intelligent decision on how to cure the problem can be made.

3. Inspect all strakes (planks) below the waterline systematically. Even if your boat is fiberglass, or unseamed plywood, aluminum, steel or ferro-concrete, now is the time to check it carefully.

   a. Any loose fasteners backed out? Missing or split plugs? Weeping spots? Loose caulking? Split plank ends? *Anything out of the ordinary?*

4. Check the garboards for rot.
5. Check the stem and scarph joint.
6. Check the condition and placement of all sacrificial zinc anodes and list those to be replaced by type and size. Evaluate how well they are doing their job. Bear in mind that their condition may be indicative of a galvanic action problem that requires further study and solutions other than renewal of the zincs.

   NOTE: Many of the checks are *not directly* associated with anti-fouling paint problems yet they are associated in many ways.

7. Check condition of propeller drive shafts, struts, stuffing box nuts, and mounting bolts. Look carefully for any evidence of pitting or unusual wear. Keep in mind that the evidence is often hidden under accumulated foreign matter.
8. Check rudder, rudder posts, swivels, all mounting hardware for condition and tightness. Have someone turn the helm inside the boat back and forth several times as you observe the action of the rudder.
9. Check cooling water intake screens. Remove and inspect them for barnacles which live inside the intake screen and choke off cooling water to the engine. Check *all* hull fittings, such as the depth sounder head, speedometer pickup.
10. Check the condition of the keel cooler (if used).
11. Check the condition of any outside ballast and fasteners. If a centerboard is used and it is possible to lower it for inspection, do it! Check the centerboard well *inside* for barnacles.
12. Check the propeller, retaining nut(s), cotter pin, shear pin, propeller guards, skeg.

    NOTE: During the off season the propeller(s) should be removed and stored for security and, if indicated, sent to the prop shop for refurbishing. Insist that they be *dynamically* rebalanced and have the shock hub renewed, the blade straightened and rebuilt by welding. Naturally, you carry a spare prop so have that checked over at the same time.

13. Check outdrive (if used).

    NOTE: The outdrive is such a complicated device that a separate checklist should be formulated using the owner's manual as a guide.

Let us repeat, this is but a sample check list. Add or delete items to suit your boat, its type of hull, and the fittings on that hull. There is no *one* best time to carry out the inspection and, except when the boat is covered with snow, almost anytime will do as long as it is well in *advance* of the fitting out period so that you will have time to analyze the results of the inspection data and formulate the rest of the maintenance plan.

TYPICAL JOB ANALYSIS

Job—apply anti-fouling paint to lower hull area. Some sample questions that should be answered in the comments section of your inspection check-off list might be:

1. Does inspection show last season's anti-fouling paint performed satisfactorily? If the answer is no or even a qualified no, what does the inspection indicate as a possible answer? Is a harder finish type paint indicated? Where can it be bought? How much? What are your choices? Does the inspection show that the paint might not have been properly applied? What should be done to correct the problem?

    NOTE: Many of the companies that manufacture anti-fouling paint have free information and instructional brochures which will describe and suggest an answer to nearly all anti-fouling paint problems. However, beware of self-annointed experts.

2. Next, think the job through step by step. Gather up all tools and materials, including those that MIGHT be needed, then prepare the surface for painting, prepare the paint (more to this than meets the eye), and finally, apply the paint.

    ### Sample List of Tools and Materials Needed for Bottom Painting Job

    Wire brushes (hand, power, coarse, and light)

    Sandpaper (coarse and medium grit)

Can of seam compound or sealant (for filling cracks, covering fastener heads)

Putty knife

Extension power cord (if power tools will be used). Be sure it is the heaviest gauge wire you can afford with grounded plugs at both ends for safety since you'll be standing on earth a lot while working.

A set of good screwdrivers or powered drivers with various sizes of blades for replacing or tightening screws.

A few wood plugs to replace any that have fallen out. Purchase them in most marine hardware stores—it is no longer necessary to buy and use a plug-cutting tool.

Hull primer paint (if the inspection indicates a start from scratch).

A quantity of anti-fouling paint designed for your type of hull, the area that you boat in, and the way you use your boat. The amount of paint to buy is governed by the area to be covered and the number of coats, and can be determined from the instructions on the can and your estimate of the hull area to be painted.

A quantity of thinner to *match* the paint you have selected. As a rule thinning is not required, but follow the directions on the can of paint if you thin.

Paint stirring sticks—or, better, stirring paddle for an electric drill.

Roll of masking tape (for waterline and boot topping). Use extreme care with masking tape as it will lift new paint as it is removed.

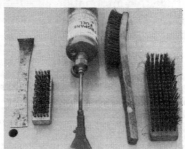

Fig. 2.9 Shown here are only some of the tools and supplies needed for a bottom painting task.

Paint buckets, throwaway liners. Old coffee cans will do nicely but tend to tip over.

Dust masks, plastic throw-away gloves, old and protective clothing.

WARNING: When wire-brushing or sanding the old coat of paint remember that anti-fouling paint contains toxic material. The dust must not be inhaled or allowed to stay on the skin. Wear a mask all the time that you are brushing or sanding and keep yourself well covered while applying the new paint. Remove any paint from the skin immediately. Plastic throw-away gloves are a good idea.

Regular three-inch brushes (such as good house brushes) or throw-away brushes. If you decide to apply the paint with a roller, which is quite acceptable, use the furry type for rough surfaces and the fuzzy types for the smoother surfaces.

And anything else you think you *might* need.

3. Then preparing the hull for anti-fouling paint, estimate the time required to sand or wire brush the old bottom paint.

Keep in mind that the old bottom paint still has a lot of use left in it if it is smooth and not chipping or flaking badly. There is no need to be overzealous in removing the old paint. In your time estimate, make allowances for unforeseen problems that sanding or wire-brushing may uncover.

Next, estimate from experience the time required to apply the anti-fouling paint, consider how many coats the hull will need, and how much surface area is to be covered. (Be sure to read through the chapter on boat painting.) Until recently, the last coat of bottom paint had to be applied just before launching so the boat went into the water with the bottom paint still wet. With the more recent products, this last-minute paint job is not necessary and your schedule will not need to take this into account. Modern anti-fouling paint must dry and set for a few hours before going into the water where the chemicals that prevent marine growth are *activated* by the salt water.

If you have done your homework, then you're now ready to begin the actual job. Try to stay with your schedule. As most boat owners know, fitting out in the beginning of the season is equal parts of socializing and actual work. Some skippers allow the socializing to get out of hand and end up having a real good time but not getting their boat in the water on schedule.

3
TOOLS, MATERIALS, AND SOME TECHNIQUES AND METHODS FOR PPM

In this chapter we shall list many (but not all) of the tools that you *might* need for a common-sense Planned Preventive Maintenance program. The list may seem extensive to you at first, but keep in mind that it is a list of suggested, not necessarily essential tools. In many cases you can either do without a particular tool or improvise and substitute your own ideas. In other instances you may be able to borrow the tools. For those bigger jobs, check with the local tool rental agency before you buy a tool. If they don't have a tool you want, often they'll get it and rent it to you.

    The first list of tools are those you will need on board your boat. The list is based on some thirty years of experience and is for the skipper who is interested in short, day-long trips rather than long-distance cruising. The list is kept to a minimum, because tools are expensive and aboard the boat they get a beating (especially if you boat in salt water). Keep the tools oiled or spray them from time to time with a preparation called "WD40" which comes in a handy spray can and is extremely useful aboard a small craft. Tools themselves often require maintenance and the more of them you have, the more you have to take care of. Although our emphasis is still on preventive, not corrective maintenance, a situation may arise where you may urgently need one or more of the listed tools — if only to lend to a nearby boat in trouble because its skipper does not have a PPM program like you do.

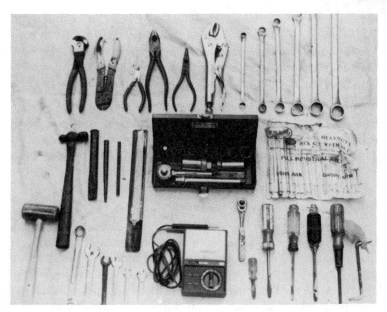

Fig. 3.1 "Typical" contents of an on-board tool box for a small craft.

TOOLS TO BE CARRIED ABOARD

Grease gun. The best type for small craft seems to be the cartridge loaded variety. It is small and easily stowed. Cartridge loading eliminates a lot of mess, and there are cartridges available loaded with a high quality of marine grease which is a *must*. There may be a few grease fittings that this type of gun cannot reach but not many. Be sure to include at least *one* spare cartridge and one or two *zerk* fittings (the little metal connectors that the grease gun connects to). From time to time the zerk fittings plug up and will not accept a charge of grease. Include a *small* container of vaseline also.

Oils. One or two quarts of regular engine oil should be transferred from cans to plastic bottles (to keep rusty rings from forming in the storage locker). Carry a can of light machine oil such as "3-in-One," a small can of any brand of penetrating oil, and a spray can of WD40 which has thousands of uses aboard a small craft.

Wrenches. A set of inexpensive, either box-end or open-end wrenches is essential. The most useful wrenches are those with a box-end and an open-end on the same wrench of the same size. Box ends are preferred as they are less likely to slip and mangle knuckles and nuts, but the open-end will fit in places the box-end won't. Keep them rolled up in oily cloth or leather.

Pliers. Many types and varieties are available, but at a minimum, you will need a pair of waterpump pliers, needle nosed pliers, diagonal cutters, and, although they have a reputation as knuckle busters, a pair of vice grips. All have an infinite number of uses aboard small craft.

Screwdrivers. Since these are easily one of the most often used tools aboard small craft, a good set of three should be carried. Try to use them *only* as screwdrivers—not as chisels, pry bars, or for testing battery connections. You will probably find that you have to carry both flat regular blade types and a couple of Phillips head types. You might consider those kits in a plastic case that have one handle for several different blades if storage is a problem.

Hammer. A small ball peen hammer seems to be the most useful although any type will do. No other tools should be used for hammering.

Punches. Try to find space for a sharp pointed punch and drift punch, which are often needed. The sharp point is for starting a drill hole in metal and for making scribing marks. The drift punch is for starting out stubborn rusted bolts and nuts, for upsetting rivets, and for clinching nails.

Adjustable wrenches. While rarely the best tool for tightening or loosening nuts or bolts, a medium-size adjustable wrench can be useful if there is room for it in the tool box.

Hacksaw. The best is one of those types with a set of interchangeable blades which insert into a holding handle. (The usual frame type of saw takes up too much storage space.)

Chisel. A small steel cutting chisel is another useful tool for stubborn rusted nuts and bolts.

Jackknife. Like the screwdrivers, a jackknife is one of the more often used tools aboard small craft. If you can find and afford one, the Swiss Army knives are excellent. Lightly oil the knife and wrap in leather or cloth.

Files. Carry one each fine-toothed and medium-toothed flat files, kept in plastic cases. Small things aboard small craft are constantly in need of sharpening or touching up with a file.

Whetstone or oil stone. Very small flat stones can be found and have a multitude of uses on small craft.

Allen wrenches. For those thousands of set screws found on small craft, a set of small allen wrenches in a plastic case are a definite need.

Socket set and ratchet. You can find an inexpensive set that includes at least a deep throat socket for the size spark plugs on your engine.

Crimping tool. This combination wire stripping, cutting, and electrical lug crimping tool is very useful in electrical maintenance. Store an assortment of crimp type terminal lugs in empty 35mm film cans with sticker labels.

Test lamp. This tool, which you can make yourself of a 12-volt bulb, a length of rubber covered so-called zip cord, and a pair of alligator clips, has got to be just about the most useful tool we know of. With it you can locate and identify most of the electrical problems that beset small craft.

Tool box. Keeping this assortment together presents a problem that can be solved with one of the new low-cost plastic fishing tackle boxes rather than the metal boxes that only add to the rust problems.

Again we want to repeat that these are only suggestions but they are based on often bitter experience. You can add to this list, but don't load up on expensive, rarely used tools such as torque wrenches, which are only needed for shore-side maintenance and repairs. Think of all the things that might go wrong with any of the equipment aboard and what tools and materials you would really need to make emergency repairs and get you safely back in port. In a bad situation, improvision (inspired by dire necessity) will often come to your rescue. I know a skipper who fashioned a perfectly functional distributor rotor out of a common paper clip that had been holding a set of charts together. Out of curiosity he ran the engine for several short trips afterwards to see how it would do and a week passed before he had to replace it with a regular plastic rotor. He now carries not one but two spare rotors.

SOME SUGGESTIONS FOR THE BOAT BOX

For small craft, the boat box may be an integral part of the tool box. It has much the same function and purpose as what our pioneer forefathers called the "possibles" box. In it were any of the small parts, materials, etc., that *might* be useful in effecting emergency repairs of every possible nature. Since a number of small, rather inconsequential things can totally disable a small craft, the replacement spares you carry make you independent of other skippers and the Coast Guard for tows and rescue. The following suggestions take into consideration all or most of those things that *might* disable your boat and cause some inconvenience if you don't accumulate and carry spares in the boat box.

Pump-alternator drive belt. If your engine uses more than one belt be sure to carry one spare for each belt. You don't go anywhere without cooling water to the engine even though the engine will run for a while without the generator-alternator functioning.

Second battery. We don't propose this as an extra, but rather as an installed back-up for the main lead-acid starting battery. This should be put aboard together with a four-position switch or one of the new dual battery isolators. Should anything happen that runs down the starting battery, a second fully charged battery will start the engine. Passengers and crew will treat electrical supplies aboard small craft just as they do ashore—as if there is an inexhaustible supply. They will leave unnecessary lights and appliances burning and rapidly run down the boat's main battery.

Cooling water hose lengths or emergency repair kit. Most marine engines have an extensive assortment of hoses to convey cooling water to various places on the engine or manifold. Some require several sizes of hose. If it is awkward to carry a spare hose length for each hose aboard you might consider our cheap but effective hose repair kit. This is nothing more than several sizes of PVC plastic hose connectors and a few flat stainless steel hose clamps. A broken or leaking hose, quickly repaired using the connectors will get you safely back to the dock.

Spark plugs. No explanation is needed here, but carry only a couple and replace the spares as you use them. Have them all gapped and ready to use.

Flashlight. In addition to the regular emergency spotlight, keep a plastic case, floatable, small flashlight in the possibles box, loaded with the longer lasting alkaline cells.

Ignition distributor spares. Any one of these parts can disable the engine completely. Since they are small and inexpensive there is no reason not to carry them in the boat box. Include a distributor cap, a distributor rotor, a point set, and condensor. Make up at least one spark plug wire using the longest lead of the plug harness as a measure.

Lamps (bulbs), fuses, and hook-up wire. Again, these are self-explanatory. Of course, not all of these can disable the boat, but a burned out running light can get you into trouble at night. When a fuse blows, try to find and remove the short circuit before you replace the fuse. You will find that a roll of plastic electrician's tape has an infinite number of uses on small craft, so add that to the spare electrical supplies.

Tube of RTV. This versatile material (RTV stands for room temperature vulcanizing) is sold under a number of brand names. It is a great sealant, can be used to plug small leaks in almost anything and can even be used to form small rubber parts. In addition, a tube (or tubes) of fast setting epoxy glue or any water-proof glue will be extremely useful.

Sandpaper. All electrical contacts require constant cleaning and burnishing aboard small craft. A couple of sheets of medium grit paper will be found useful.

Gasket paper. It will be impossible to carry all of the pre-cut and formed gaskets that you will need from time to time on small craft. It is therefore much easier to carry a sheet of gasket paper and gasket cement. A leaking manifold joint or oil leak can be repaired with a new gasket cut from the paper using the old gasket as a template.

Sail needles, sail palm, heavy thread, and beeswax. While hardly essential to the skipper of a motorboat, the sailboat skipper will find these a must.

Assorted cotter pins. Make sure these are rust-resistant metal pins. Only a few need be carried. Whenever a cotter pin is removed, don't try to reuse it, but put in a new one—they're cheap.

Fasteners. Carry screws, nuts, bolts, flat and lock washers, nails and any other assorted fasteners you can think of. Make sure these are rust-resistant and avoid brass and galvanized types which are both susceptible to galvanic action and corrosion.

Nylon twine. A small spool is all you need. It has many uses in the hands of an inspired improvisor.

Windshield wiper blade. Ever try piloting a boat in heavy water or hard rain with the old wiper ravaged by sun and sea rot? Carry a spare and replace the spare the minute you use it.

Stove fuel. Carry the smallest amount you can tolerate. It is a fire hazard and you can always eat and drink cold if you must.

Air cans. If you carry an emergency horn powered by air or gas-charged cans be sure to carry a couple of spare, charged cans. The Coast Guard insists that spares be carried for this type of horn.

Dry cells. Inventory all the equipment aboard that uses dry cell batteries for power. The most critical (weather radio, radio direction finder, etc.) should have a set of spare alkaline type batteries which should be stored in their original packages to keep the salt air away from them. Renew the spares each season, or repower with the spares at the beginning of the season and restock the spares. The shelf life (time the batteries will still be usable after storage) cannot always be trusted.

First aid kit. Carry and store the best that you can afford. Most of the made-up kits are a cruel joke. We recommend one that includes the new plastic devices for mouth-to-mouth breathing assistance, and a second implement that helps people choking on food bits (neither of these is of any use unless you know *how* to use them properly). Lay in supplies for burns, cuts, and hangovers.

TOOLS FOR PPM AND CORRECTIVE MAINTENANCE

In the following paragraphs we'll list and describe some (not all) of the tools that will be needed in a PPM program. Most of the tools listed are for woodworking since many boats are still made of wood. However, most of the tools normally reserved for wood can also be used on fiberglass. A few (very few) extra tools are required for aluminum hulls. Our list is not intended to cover *all* the tools that *might* be required, just those that are most often and commonly used. The average homeowner will already have most of the tools and as a rule will not have to buy any new ones.

Many other tools that are useful may make the job go quicker or reduce the overall labor but are of a one-time use nature. For these, you should consider either borrowing or renting. A good idea is to line up two or three other skippers with the same requirement for a special tool and rent or buy together. Again, these are suggestions only and you should make your own decisions on what tools you should have. As a general rule, if you must buy the tools, by all means get the best you can afford but don't go overboard on industrial-professional grade tools. The tasks of PPM cannot justify the high expense of that grade of tool unless you get so good at PPM that you go into the boat maintenance business for yourself.

Hand Tools

Hammer. You should already have a small ball peen hammer in the boat box which can be used for peening over rivets and upsetting other fasteners. A good claw hammer may not be the best tool for a particular job (a mallet is for working with a chisel or caulking a plank) but it will do for all but a major caulking job.

Saws. A good crosscut saw with eight to ten teeth to the inch should cut fine enough for even the most fancy work. A rip saw is for pros and cannot be justified these days. Smaller saws such as a fine-toothed hacksaw or a coping (jig) saw all have their uses but don't buy them unless the job calls for it.

Planes. A medium (ten-inch or so) jack plane and a small block plane will cover ninety percent of what you need to do.

Drills. A bit brace is a must — it is too useful not to have the best you can afford, that is, one with a reversible ratchet, and a ball bearing chuck and head. To go with the bit brace you will need a set of wood bits and an expansion bit for the bigger holes. A rose countersinking tool is needed when a flat head wood screw must be driven below the surface of the wood so the head may be covered. And, most important, have a set of screwdriver bits both for flat head screws and for Phillips head screws in as many sizes as you think you can afford.

Hand drill. Include only if you do not plan to have an electric powered drill.

Chisels. A set that includes a ½", a 1", and a 2" blade should do nicely. Keep these not only sharp, but honed, and store them in a leather case.

Screwdrivers. Those in the boat box are good enough for the job. A push-pull (or so-called "Yankee") screwdriver takes a lot of work out of job that involves inserting many screws into wood, such as laying a new deck. However, if you have the bit brace it is even better for the job.

A try square with leveling bubble and a sliding bevel. Since just about all woodwork in a boat seems to have a bevel of some kind on it you can consider the sliding bevel as nearly essential.

Clamps. You will need at least two pipe clamps and as many of the C-clamps as you can lay your hands on. Again, don't go overboard—try to borrow or rent for a big job.

Power Tools

Power tools take the drudgery out of small craft maintenance, so give careful consideration to each of these. None of the larger shop tools are listed as they are not needed and that kind of work usually can be done at the lumber yard where you buy your wood.

Electric drill. Because of its great versatility you should have the best electric drill you can afford. These tools range from the inexpensive intermittent service drills to the industrial grade costing well over $100. They can be run all day for years without failure. Cordless electric drills must be recharged after short use, and unless there is a lack of power available, they are not

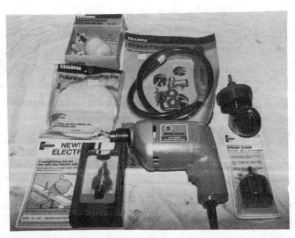

Fig. 3.2 A modern electric drill with only a few of the accessories available to the small craft owner.

recommended. If possible, get one of the new variable speed drills. With a little practice you can even learn to drive screws without other attachments. For any large job such as the repair and replacement of a garboard plank or a new deck job where extensive drilling and screws are involved, consider a low-speed chuck attachment for driving the wood screws.

Drill attachments. You will need a good variety of twist drill bits to fit the chuck of your electric drill. A tool that drills a pilot hole, a counter sink, and a hole for a wooden plug, all in one operation is both inexpensive and very useful as a step-saver. A must for fiberglass boats is a rubber disk to use with buffing cloths and pads. Also needed is a second rubber disk attachment to which sanding discs can be glued. Rotary wire brushes come in all sizes from fine to coarse wire—super great for rust fighting and burnishing metal clean and bright. Rotary hole cutting saws are handy for cutting perfectly shaped and sized holes. Shaping tools are what you need to gouge out wood to form complex shapes. Don't forget a grinding stone that will chuck in an electric drill. You may not need any of these attachments, but they are available and should be considered. If you're smart, you'll put such items on your Christmas and birthday lists.

Sabre saw. Like the electric drill, this is a very versatile tool. A phenomenal number of different blades are available which can be changed in seconds to cut just about every kind of material found on a boat. With a little practice a sabre saw can be made to do all the things the other saws do—perhaps not as fast but

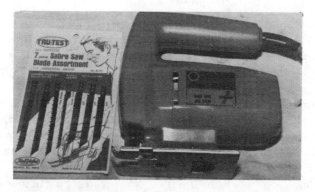

Fig. 3.3 Another versatile power tool—the sabre saw.

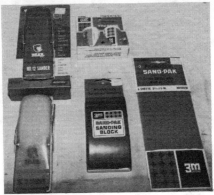

Fig. 3.4 Examples of sanding devices.

just as well. The better saws come with variable speeds and with edge and circle cutting guides.

Power sanders. These tools are great but a luxury that you can do without. Besides, if you have the electric drill with the right attachments you have a power disc sander anyway. If a big sanding job is coming up, consider renting or borrowing a belt sander, a large disc sander, or an orbital sander. Use a belt sander for flat surfaces only and use it with great care as it can gouge out a lot of wood before you realize what's happening. The orbital sander can be used on the side of a boat, but that is a tiresome job. For finish work the orbital sander and disc sander will leave marks in the wood. No machine can compete with hand sanding for a truly beautiful finish and if you are particular about your brightwork, you'll be doing a lot of hand sanding.

MATERIALS FOR USE IN PPM PROGRAMS

Grease, oil, and general lubricants should be selected for marine application. These products must cost more, but, in general, they will stand the gaff better. Most of the silicon lubricants stand up well, and one of the most useful is a spray can of WD40.

Paint, varnish, and wax, like the lubricants, must be chosen for marine application. They are formulated for that purpose and even the finest grade of housepaint might not last a week on a boat. Many of the better grades of paints and varnishes contain the necessary elements—ultra-violet resistant or absorbant chemicals. You might be surprised to know that the gel coat on fiberglass boats is very porous and for that reason special waxes have been developed with this and the oxidation and UV damage in mind. Wax that is great for automative acrylics and lacquers might damage the gel finish on your boat. Select and use waxes that have been formulated for *boat* fiberglass gel coats and nothing else. There are some fantastic paint removers that are safe for the do-it-yourself skipper to handle. If you use them, remember that paint removers could attack the resins that hold your fiberglass boat together. Rust removing jellies will take the rust off of not only ferrous metal, but used with care, will remove those ugly stains on the bow of boats that accumulate unsightly "mustaches" from boating in polluted water.

Wood for PPM

When selecting wood for either preventive or corrective maintenance work (which will usually be the case), take the time to pick out the best you can get. Look for wood with parallel grain, knot free, no pitch, no checks or splits, and be ready to pay extra for dry, well seasoned wood. When repairing or replacing plywood get marine grade—don't accept "exterior" grade as equal because it's not! Exterior grade has unseen voids in the internal plys and those gaps are potential wood rot pockets *just waiting for a bug or spoor to get started.* For ribs, floor supports, stringers—any part that must bear weight and stress—select well seasoned *white oak.* For planking, *Eastern white cedar* is best, but be wary of the many other less durable cedar woods from all over the country. The oddly named *Hackmatack* wood makes the best natural grown knees for supporting transoms, thwarts, stems and the like. It is tough, durable, and rot resistant. *Mahogany* cannot be equalled for brightwork and is justly called millionaires' planking. *Honduras mahogany* is the best but Philippine mahogany isn't really a mahogany at all. Of all the wood available for boats, teak has got to be the best but it will seem that platinum could be used for the same price. In addition to its cost, teak is hard on tools but it is by far the most durable of boat woods as it is nearly impervious to rot because of the chemical characteristics of this wonder wood. Teak does not need paint or finish of any kind except for an occasional oiling.

Bonding agents (glue). Here's where we boat keepers have finally got a break. There are many truly great formulations which (when correctly used) make anything adhere to anything else. One or two are so efficient that you must be careful not to get it on your fingers—if it happens the fingers can only be separated very carefully with a razor blade. Some bonding agents, especially the epoxies, are two-part systems—that is, the glue comes with a separate activating agent which is added only when the glue is to be used. Many glues are limited by their *pot life,* which means the glue starts setting up immediately as the activating agent is mixed. The setting-up time varies from product to product, so be sure to evaluate the job before you select a glue. Some of these esoteric bonding agents are *thermal setting*, meaning ambient heat is used to make them cure and harden, and therefore cannot be used below certain temperatures. Others cure and are hardened by the moisture in the air. Because of this feature, these types are best for a bonding job where the finished bond might be wet all or most of the time. Other new materials flow into and fill cracks between planks of the hull or deck and turn into a rubber that flexes with the hull.

Fiberglass, plastics, and resins

"If God had meant us to have fiberglass boats he would have created fiberglass trees." Only a hardheaded traditionalist would fail to see the humor in this. Few can fault the advantages in the use of fiberglass for boats. A boat with a well maintained fiberglass hull is a thing of beauty and a joy for several years. One recent phenomena is that some—not all—fiberglass boats have been sold for more money than they originally cost. The reasons for appreciation in value are complex and primarily based broadly on a lot of world-wide economics (much of the basic raw material for fiberglass derives from petroleum crude oil). As a result, a ten-thousand-dollar two-year old fiberglass boat might well go for twelve thousand dollars today. In spite of the claims for fiberglass hulls, they are *not* maintenance free. True, they do not require the same care as a wood or metal hull, but as any fiberglass boat skipper will privately admit, some maintenance is required. There is a substantial amount of choice in the market for fiberglass maintenance materials, but be careful about bargains. For example, some fiberglass cloth not made for marine use will still have a chemical sizing in it and the epoxy resins will not stick to it. Be sure that anything you purchase for either PPM or corrective maintenance is for marine use. In chapters to come we will further detail the use of fiberglass materials in maintenance.

Wood preservatives. You should not put any bare wood aboard a boat. After a piece of wood has been cut and fitted it must be soaked with a good brand of wood preservative *before* it is installed—this means all sides and ends of every stick. It has been found that when the preservative is applied *before* it has been cut and shaped, the preservative tends to make the wood hard to work with and dulls the tools. So cut and shape first, soak with preservative second, install and paint third. Most paint manufacturers market materials designed to fight wood rot. Most, if not all, of these products are based on copper napthanate mixed with the usual wetting agents (to make it soak into the wood better). Wood well soaked in these preparations will normally fight off the ravages of rot and it is truly the best method of wood rot prevention. Some caution is required when using preservatives because certain kinds of paints will not adhere to a surface that has been protected with such preparations. Be sure to read the directions before using these preparations. A colorless form of preservative is available for use on mahogany or other woods that are to be varnished, but staining the treated wood may be a problem. As always, read the directions.

SOME GENERAL COMBAT TECHNIQUES AND STRATEGIES AGAINST THE ENEMIES OF BOATS

Fighting the Ravages of Sunlight

Whenever possible moor and store your boat under cover. If you are one of those truly fortunate skippers with a private dock,

Fig. 3.5 Example of a small craft mooring cover that protects boat's finish from weather and ultra-violet rays of the sun.

Fig. 3.6 Although expensive, this provides the best mooring protection for small craft.

consider erecting a roof of some sort over the dock. The value of this is apparent in the condition of some of those old mahogany hull lake boats that were kept in a boat house of some type. Protected from the ravages of ultra-violet radiation, many of these boats are seen today at antique boat shows in perfect condition. When a roof is not possible, an investment in a mooring cover will pay off. At the end of a day's boating, take time to wash down the boat to get rid of the dried salt crystals and button up the boat in its mooring cover. Thieves and vandals will be less likely to invade a covered boat; they prefer an open boat they can get into quickly. In addition to a mooring cover, always choose paints and varnishes that contain UV resistant or absorbent chemicals. This is especially important for the cabin roof, deck areas and any exposed brightwork in the cockpit. Don't expect perfection in paints and varnishes because it hasn't been reached yet, although every year they get better.

Fighting the Ravages of Air (Oxidation)

It seems an oversimplification, but a well kept PPM program of paint, varnish, and wax is most effective in keeping the oxygen in the air (and water) from combining with the structures in your boat to cause oxidation. A freshly painted hull, well dried and set, given a coat of wax and buffed before it is launched for the season, will help in many ways. First, it will have that super glossy sheen. Second, this serves to partially insulate the surface of the paint from both air and UV.

On fiberglass, protection against air is a necessity. Freshly painted wood boats may look so beautiful that waxing does not seem necessary. But wax anyway—it will pay off next year when you find that the hull does not require painting again, just rewaxing.

Protecting metal surfaces against air is a little more complex but the principles are the same—that is, some kind of a barrier must be put between the air and the metal. This barrier can be paint, plastic, chrome plate, gold plate or what have you. Since metal, for the most part, is not porous, it requires different paints and surface preparation to get the protective coat to stick. Preparation involves removing *everything* that isn't base metal. The biggest problems are vestiges of wax, oil, or grease in any form which will prevent even the highest quality paint from adhering to the metal. Therefore, begin by wire brushing, sanding, or scraping away all old finish or rust until the metal is clean and shiny. Then wash it with a strong detergent. Next, wash the metal with a degreasing liquid such as acetone. After the metal surface is clean and degreased, do not touch it with bare hands as there is enough natural skin oil on your hands to prevent the paint from sticking to the metal. If the metal is aluminum you will have to use an etching chemical to rough up the surface so the primer and finish coats will stick. Other metals require a metallic based priming undercoat, such as zinc chromate, (or wash primer) or one of the new epoxy undercoaters. Finally you can begin painting the finish coats on in one to three layers. If a small area is to be protected, consider using the spray cans of paints and undercoaters that are designed for this purpose. If the metal to be protected will be subjected to extremes of heat (such as the manifolds on an engine), there are special heat-resistant paints for that purpose, but don't expect them to last—they cannot.

Fighting the Ravages of Salt

Most of the tactics for protecting against salt are similar to those for protecting against air and involve the upkeep and maintenance of protective coats of paint, varnish, and waxes. Since salt is an active substance it will, when deposited on other substances, try to form chemical combinations which are rarely beneficial. Salt is the most frequent cause (in combination with air) of metallic corrosion. Skippers who operate boats in fresh water, even far from the ocean, must guard against the ravages of salt. Highways are salted to melt snow and ice and if you move your boat on a trailer over a salted road you can anticipate big trouble for both trailer and boat. Even in the spring, after a rain, there is still enough salt on the road to start corrosive action on boat and trailer. If your boat is an aluminum hull, then there is added danger from salt. While fresh water does not neutralize salt it will dilute and dissolve the deposited crystals and carry them away. Wash down after any trip on the road and wash down with fresh water after every ocean trip. Don't throw that old garden hose away. Splice the leaks with cheap couplers and take it down to the dock to use for washdowns. If you swing on the mooring, try to wash down during fueling as often as possible. Salt in the atmosphere dearly loves the copper used in electrical connections. It will help in forming that green coating (verdigris) on all uncoated copper surfaces. Verdigris is a poor electrical conductor and since small craft in general use low voltage electrical systems, the smallest unwanted resistance to current flow cannot be tolerated. A lot of preventive maintenance time is taken up by this problem.

Fighting the Ravages of Galvanic Action

If you know what galvanic action is and the conditions that must be present for it to exist, then a large portion of the battle is won. When two different metals are immersed in a liquid that can conduct an electric current (salt water) an electric difference of potential (voltage) will exist between the metals. Due to this difference, an electric current will flow from the metal that is less noble (active) to the more noble (inactive) metal. Make doubly sure that any new electrical equipment installed on the boat is connected with *exactly* the same electric polarity as *all* the other equipment. If you do not do this, you may be inadvertantly setting up one of the conditions for galvanic action. One of the best defenses is the installation of a heavy copper ground strap that runs along the keel, INSIDE the boat. Every piece of electrical equipment is grounded (attached electrically) to this strap placing *everything* aboard at the same electrical potential. Normal wiring, however, carries the electrical current, while the ground buss performs anti-galvanic action and electrical safety. As a last, expensive, but effective device against galvanic action, there is available an electronic device which automatically monitors galvanic action and supplies countercurrent and voltage to stop it. Note that galvanic action can come from the least expected quarters.

Fighting the Ravages of Rust

When free oxygen in the air combines with atoms of iron, ferric oxide—rust—is formed. It may be that the main objection to rust is esthetic and it does not necessarily follow that the *action* of rust alone is destructive, but when accompanied by corrosion and/or galvanic action there is real trouble brewing. Oddly enough, the best protection against rust is a coating of rust. Once a layer of rust has formed on the surface of a ferric metal, further formation is limited. Fighting rust usually takes the form of coating the susceptible metal(s) with preservative paints specifically designed for that purpose. Apply light coatings of grease when paint cannot be used or, where possible, substitute another metal that does not rust. Rust can be removed from metal by vigorous wire brushing and sanding with crocus cloth. The bare metal is then washed with a degreasing liquid, primed with a metal primer, and painted. Rust can also be removed by a substance called "Naval Jelly" which is sold under a number of brand names. A rust spot or streak on fiberglass or paint can be removed by rubbing with a lemon cut in half.

Fighting the Ravages of Wood Rot

Again, half the battle is knowing the conditions and the cause of the problem. Fresh water, either from rain, condensation, or wash downs, allowed to stand can start wood rot. Lack of ventilation in dark damp places, unprotected bare wood, and an infection from air- or water-carried spories and bacteria combine to cause the problem. Fortunately the presence of wood rot can be detected by its unpleasant odor. If you detect it on your boat,

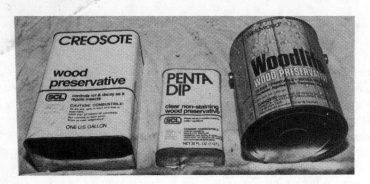

Fig. 3.7 Wood hull rot fighter — a must for the wooden boat owner.

don't panic. Find it, exorcise it surgically and replace the damaged wood with fresh wood SOAKED in a reliable preservative. This is corrective maintenance not preventative. PPM calls for regular inspection and resoaking bare wood in preservatives. Controlling leaks and drips from fresh water and maintaining good ventilation within the boat (especially when stored in the off season) are sound and effective preventive tactics.

Fighting the Ravages of Plants and Animals

Primarily this involves the selection, application, and maintenance of anti-fouling paint appropriate for your boat, your boating area, and your boating conditions. We will discuss this in greater detail in the chapter on painting.

Fighting the Ravages of Man

The most serious threats are laziness, failure to follow instructions, inability or reluctance to learn new skills (or *EVEN TO TRY*), and carelessness. The best defense is to take boating safety courses and practice and polish your boat-handling and piloting skills. Encourage your shipmates to do the same.

4
PAINTING: THE FIRST LINE OF DEFENSE

In the never ending defensive war against the enemies of small craft, painting is the first line of defense. You get the best return for time, money, and effort expended. There is more to painting than you might think. Painting is very final and a bad job can be corrected only with the greatest difficulty. It is essential to do it right the first time. In the following paragraphs we shall present a systematic approach to a PPM painting program. Fortunately, we have access to a lot of new products, labor saving tools, and effective techniques, and more and better are being developed. Nearly all these products are available to the average boat owner and not restricted for use only by the professionals.

PLANNING FOR PAINTING

Little else affects the overall appearance of small craft so much as its paint color scheme. If you are satisfied with your boat's color scheme, then skip this discussion. If not, consider the following factors. First, using a photo of your boat make a sketch of it on white paper. A general outline of the boat will suffice. Now, try a little experimenting with crayons or markers, starting with the topsides. Remember that white requires more care and is not a particularly exhilarating color. However, white or light colors tend to make a boat look larger while dark shades make it seem smaller. Once you have selected the topside color, you can think about a deck color. The right color will bring out the best lines of your boat. Keep in mind that the lighter the

64 Painting: The first line of defense

Fig. 4.1 Solid black hull gives a "blocky" appearance to the boat.

Fig. 4.2 This white forward deck area reflects undesirable glare.

Fig. 4.3 Dark decks with light topsides give a boat that "top-heavy" look.

Planning for painting 65

Fig. 4.4 A plain white hull tends to make a boat look lower and longer.

Fig. 4.5 Good balance in colors.

color, the more it will reflect the heat of the sun and the cooler it will be underfoot and in the cabin. Dark colors, like dark clothing, are warmer and tend to hold the heat. As captain, pilot, and helmsman you will appreciate looking out over a forward deck that is greyed down to diminish reflection of the sun into your eyes. Dark decks with white topsides give some boats a top-heavy look (thereby raising some doubts with your passengers). If you are dubious, try it out on the sketch with crayons. On the hull, plain white will make it look lower and longer while dark colors will tend to make it look high and bulky. The superstructure is next and if it has a trunk cabin or raised deck exterior and it seems too high to you, it can be brought (seemingly) down by painting with a light color. Try it out on the sketch before you make up your mind. We are no longer limited to the old red bottom paint. Color is not considered one of the essential practical features of bottom paint. Since it is available in many shades at no extra cost, choose the color bottom paint that pleases you. If yours is a sailboat there will be times on a windy reach when your bottom will be showing, and it should look as nice as the rest of the boat. Choice of a snappy boot topping will achieve the desired total effect. Boot topping is an ornamental stripe, usually a darker contrast to the hull, painted between the water line and the higher section of the hull. While it may appear to be no more functional than a racing stripe on a car, the boot topping serves to disguise the dirt and filth from polluted waters. Some combinations you might try are green or white boot top with red bottom, or red or white boot top with green bottom, red, green, white, or blue boot top with bronze bottom.

For cabin interiors stay with a light tint of the deck color. Choose colors that are soft, restful and pleasing to the eye. The same principles hold true for the open cockpit.

PAINT SELECTION BY TYPE (FORMULATION)

Most new marine paint formulations fall into one of four general categories based largely on the resin used in the paint. Be warned that a paint that serves well for one job might be inappropriate for a different application. The table below summarizes these

Type	Use	Advantages	Disadvantages
Alkyds	Anywhere above the water.	Greatest color variety. Quick drying. Inexpensive.	May not adhere to fiberglass. Not durable.
Vinyls	Most used for bottom paint.	Flexible and tough. Best for aluminum.	Might lift paints other than vinyls.
Polyurethanes	Above the water. Excellent for areas exposed to salt, sun, and pollution.	Tough. Good in warm water.	Expensive.
Epoxies	Any surface.	Really sticks Tough and durable. Perfect for concrete and metal.	Won't hold color well. Expensive.

paints by type and use, together with the advantages and disadvantages.

Whatever type paint you choose, stick to that type unless the maker says otherwise.

ESTIMATING YOUR PAINT NEEDS

It is safe to assume that a gallon of paint will cover 500 square feet. However, if it is new wood or wood that has been stripped of old paint, you can figure that a gallon will cover 325 square feet. For varnish, about 750 square feet is a reasonable expectation on recoats and 500 square feet on new wood. To help you more accurately estimate the exact quantity, you will find the following arithmetic useful.

Cabins or deck houses. Multiply the height of the deck house in feet by the distance around (girth). Subtract the area of any openings, such as ports and hatches. If the deck house is to be painted, divide the result by 325 for the priming coat, and 500 for each finishing coat. Divide by the varnish numbers if it is to be varnished.

Decks. Multiply the length of the boat in feet by its greatest beam width and multiply the result by 0.75. Subtract the area of the cabin houses, hatches and ports. Divide the remainder by 325 to get the gallons or fraction of gallons needed for a priming coat, or by 500 for each of the finish coats. A calculator will help!

Hull or topsides. Multiply the overall length in feet by the greatest freeboard (distance up to the rail from the waterline). Multiply the result by 1.5 and divide by 325 for new work primer, and by 500 for each coat of finish to get the gallons or fractions of gallons needed.

Bottom. Multiply the waterline length by the draft (greatest distance from the water line to lowest point under the boat) in feet. For a keel boat, multiply by 3.5, and multiply by 3 for a centerboard boat. Divide the final result by 300 for priming needs and by 400 for each coat of bottom paint needed.

Using the estimating formulas to keep from buying more paint than you really need is your first opportunity to economize with the PPM program. Once opened and used, modern paints do not store well. Unopened cans will usually be accepted back by the dealer for credit. Don't be afraid to ask for advice from a dealer who knows marine paints. Ask for and take any free booklets and brochures that are often available from the paint manufacturers. The master painter in the local yacht yard is also a good source of advice.

In summary, select your boat paint first by the suggested color scheme, second by the type (formulation) you'd like to use, third by your boat hull—wood, fiberglass, or metal—and finally by the use that your boat will have and the area where it will be used. Seek advice of the experts either in person or through their booklets. Estimate your paint need by amount, then shop around for the best deal.

GETTING READY FOR PAINTING: TOOLS

Wire brushes. You may need *wire brushes* and *scrapers* for getting the surface ready to paint. Wire brushes come in every shape and size from very coarse to quite fine. Choose those with stainless steel bristles even though they are more expensive. (They don't leave broken bristles sticking in the wood or fiberglass to rust and bleed all over your nice new paint job.)

Scrapers. Scrapers also come in all shapes and sizes. The most useful is the hook type scraper which comes with replaceable and interchangeable blades. There's a knack to using a scraper and success depends on how well the edge of the scraper is sharpened. Keep a flat file on hand while you scrape. As soon as the scraper stops raking off the undesired material, stop, brace the scraper against something, and give the edge several strokes with the file, keeping the file at an angle the same as the original edge. The hook scraper is sharpened only on one side to form a razor sharp hook-like edge. To save its cutting teeth, stroke with the file *only* in the forward direction, *never* draw the file backwards.

Sandpaper. These days there is no sand in good sandpaper—the grit might be silicon carbide or aluminum oxide. It goes by a lot of other names, such as production paper, crocus cloth, wet and dry, and garnet paper. Whatever the grit material, sanding material comes in a vast variety of forms—sheets, discs, belts, and so on. The basic three will be most useful for maintenance jobs on small craft. What you need are the production papers, belts, or discs that consist of an abrasive of aluminum oxide, silicon carbide, or garnet paper. Abrasive materials vary in degree of hardness. Generally, the

Getting ready for painting: Tools 69

Fig. 4.6 Examples of wire brushes.

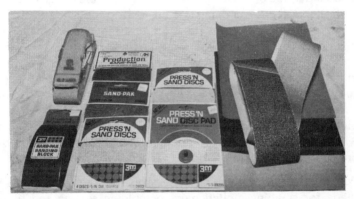

Fig. 4.7 Sanding equipment most commonly used.

harder the abrasive grit, the better it cuts, the greater it costs, but the longer it lasts. Although technically a misnomer, for convenience we will use the common term "sandpaper" in our discussion here.

Abrasive papers are now reliably rated by a mesh number which indicates the *size* of the particles of grit stuck to the paper. For example, a size 80 grit will have a 0 grade number and can be considered a medium paper. A 120 grit, grade 3/0 is rated fine, and a 30 grit, 2½ grade is an extra coarse paper that might be used in removing paint.

Some hints on using "sandpaper"

Buy the hardest grit paper because it lasts longer. If you wish, use one of the many kinds of wood and plastic holders. The best papers "load up" during use—that is, the spaces between pieces of grit become packed with cut material and the paper *seems* to have worn out. Before you throw it away try tapping the paper on a flat surface or use a wire brush on it. You'll find it will continue to cut well. Save used paper and use it as a finer finishing paper. For fiberglass or its gel coat, the best paper is the wet and dry type which you use *wet*. As the plastics are cut by the paper, they rapidly "load" up the grit. Keeping it wet will inhibit the loading effect. Sanding discs for power sanders will make the job easier but must be used very carefully because the circular marks left in the material are difficult to get out and will show through the finish coats of paints or varnish. On flat horizontal surfaces nothing is equal to the belt sander for fast action. Yet, like the disc sander, the belt must be used with great care because it cuts so fast that there is a danger of difficult to remove gouges. All sanding is ninety percent drudgery and ten percent skill. The main criteria is how well you want the finished job to look. Sand *only* with the grain of the wood, *never* across the grain (cross-grain sanding leaves scratches). Start with a grit size that seems too small to do the job, and work up to a size that cuts the way you want. Don't use either crocus or emery cloth for any electrical burnishing. The grit in both is metal and a conductor which might cause trouble later on. Finally, if the job is big enough, consider renting a power sander of your choice.

Putty knives. Have at least two sizes, one with a very wide blade for scraping melted paint or holding sealing and filling material as you work with the knife with the smaller blade. Clean

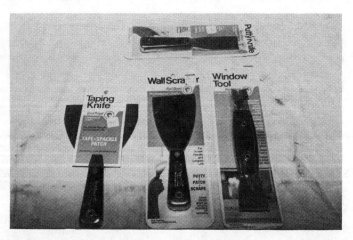

Fig. 4.8 Tools and materials for paint removal.

after using while the material that was spread with the knife is still soft. Some of the new sealers, resins, and glues are too hard to clean off after setting.

Tackcloth. A tackcloth is a must for any kind of brightwork varnishing job. After the surface has been finish sanded, only a tackcloth will remove the remaining dust. A tackcloth is easy to make and keep. Buy some cheese cloth or use an old towel, wet with water then sprinkle with turpentine. Wring until the turpentine spreads throughout the cloth. Finally, drip varnish all over the cloth and wring it again. Keep the cloth stored in a sealed can and renew it from time to time with more turp and varnish.

Tools for paint removal. Since there are many strong opinions and much controversy, it is best to conduct small experiments to determine what tools and methods for paint removal suit you and your boat. In general, there are three methods for taking paint off of a surface. There is no one best method for all jobs. The material under the paint and the paint type itself are governing factors. Paint can be removed by straight sanding and scraping, which are at best slow and tedious work. Also, with handsanding you maintain control. Paint removing chemicals are effective and will, when properly used, take paint off with the least damage to the under surface. And, since the Occupational Safety and Health Agency of the Federal Government rulings, most of the toxic chemicals in paint removers have been

Fig. 4.9 Painting tools, give them the care they deserve.

eliminated. The new and safer chemicals are just as effective in removing stubborn paints as the old—sometimes even better and even easier to use. One brand comes as an inexpensive powder which is mixed with water to a desired consistency of paste. This stuff will stick to the sides of a boat and depending on the surrounding temperature will lift the paint from the hull like a lobster shedding his skin. However, there is still some form of caustic involved and unprotected skin can be stung.

Heat is another method and sometimes is faster than chemical paint removers, but heat cannot be used on fiberglass or aluminum hulls. The old gasoline blow torch that was once used for paint removal might have given the heat method its bad reputation. However, the newer propane gas torches with the flame spreaders seem to work quite well. The idea is to direct the flame at the paint surface, keeping the flame moving at all times. Heat the surface *only* enough to *soften* the paint, *not* till it bubbles, scorches, and smokes. Follow the torch flame with a wide bladed scraper or sharp putty knife, being careful not to gouge the wood. Use the torch to get the heaviest layers of paint off so that you can finish by sanding off the balance. Other heat sources are electric irons and infra-red heat lamps. While slower, they are safer in that neither one uses an open flame. Remember that heat, used in paint removing, makes the wood brittle and raises the grain when it is not used correctly. In any event, make up your *own* mind as to the best method for you and your boat.

When the hull is aluminum, neither chemicals nor heat can be used to remove old paint. The chemicals may react badly with the metal and the heat can change the temper of the metal or burn right through. Carefully sand off the old paint with a fine grit *non-metallic* paper. If you use a metallic grit, pieces of the grit might become imbedded in the aluminum and set up a galvanic action pocket that eventually will eat a hole right through the hull!

Paint brushes. Like so many things, brushes come in a wide variety of types, sizes, and costs. In general, buy the best nylon bristle brushes that you can afford. They may be even better than the hard-to-find natural bristle brushes. For a PPM program you should have at least several brushes—a small fine cutting-in brush for fancy work, a sharp pointed (like a chisel) one-inch brush for varnish work (use your varnish brush for nothing else), a two- or three-inch brush for decks and hulls and a heavy flat or round brush for applying bottom paint. Throwaway brushes are

Getting ready for painting: Tools 73

Fig. 4.10 You'll find a bucket a more workable receptacle than a paint can.

okay but they will not hold or spread the paint well and they will drip. Use a good brush and clean it immediately after the job. Proper cleaning will save money, time, and effort, and the brush will do a better job next time. Between coats, keep the brush suspended in a square can filled with equal parts of turpentine or thinner and linseed oil. Stored this way, it won't be necessary to clean the brush between coats, just *wipe* it on scraps of clean wood (don't pound it) before you start the next coat. Some brushes come with a storing hole already drilled in the handle. Fill the storage can with just enough keeping mixture to cover the bristles. Don't store the brush in water, which will damage the metal and swell up the wood of the handle. If the storage can is too much trouble, try keeping the brush overnight in an airtight plastic food bag tightly closed with an elastic band. If the paint you are using is one of the two-part epoxies which sets, not dries, then you will be forced to clean the brush completely between coats. Otherwise you will be able to use the brush for a hammer after it sets overnight. In any event, if you have your brush cleaning tools ready—thinner, brush cleaning liquid, metal brush comb, or wire brush—and clean the used brushes right away, they will be good as new.

Most skippers paint so frantically that they are too tired to face the brush cleaning job at the end of the day. Schedule brush cleaning time in your job analysis.

Paint buckets. These are more essential than you might think. When you are stirring up a new can of paint the best method is to

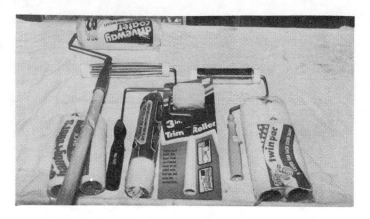

Fig. 4.11 A wide variety of sizes and shapes make rollers more applicable to small craft painting tasks.

pour about half of the thin unstirred paint off the top into the paint bucket, then, as you stir the thick gooey stuff at the bottom, gradually add small amounts of the thin mix from the paint bucket to the paint can. A good paint bucket is lower and has a wider bottom than a paint can or an old coffee can—thus, it is less likely to tip over. When painting, pour only enough paint into the paint bucket to cover the *first one-third* of the brush. This method prevents overloading the brush *and*, by keeping the paint can covered, you prevent the escape of the more volatile thinners and drying agents in the paint. There are inexpensive plastic paint buckets available. Other buckets have disposable liners which help with cleaning up after painting.

If you are going to varnish, consider a varnish cup. This is no more than a cheap tin or plastic cup in which you drill a couple of holes near the top edge. Cut a straight section of wire out of an old coat hanger and insert it through the holes. Bend the ends of the wire over so that it can't fall out and you have one of the handiest gadgets for laying up varnish. After it has been loaded with varnish draw the brush lightly over the wire to help stop bubbles in the brush. By pouring off varnish into the cup you accomplish the same results as with the use of the paint bucket.

Rollers. If you do decide to try rolling the bottom paint (which is far easier and perfectly acceptable), be sure to get the furry roller that will hold and spread a lot of paint. Remember that the active ingredient in bottom paint is metal which quickly sinks to

Getting ready for painting: Tools 75

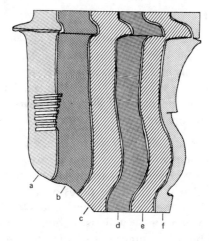

Fig. 4.12 Properly painted outdrive has (a) wash primer coat, (b) barrier coat, (c) first finish coat, (d) 2nd barrier coat, (e) and (f) final finish coats.

the bottom of the can, so stir the paint each time you load the roller tray. When you shop for the roller handle, look for the type that has a heavy threaded hole into which you can attach an old mop, swab, or brush handle with the same threads. You now have a roller with an extender pole that will, in many cases, help to reach hard-to-get-at places under the bottom of the boat without covering yourself with paint. To save yourself time and trouble put the roller paint tray inside a plastic garbage bag. When you are through for the day, dump back the remaining paint, remove and discard the plastic bag and the paint tray is clean and ready for the next go-round.

Leftover paint. If you have done your estimating well, there will be very little leftover paint, which is best given or thrown away. At best it does not store well and anything other than a water base paint is a fire hazard and a storage problem. However, if you must store a partly used can of paint try pouring a little thinner or turp on the top of the can before you seal it carefully. Having sealed the can, mark the date purchased and the purpose it was used for on the top with a felt tip pen. Turn the can *upside down* and store it that way in a dry place where it won't rust or be subjected to temperature extremes. By storing it upside down that thick skin of paint will *not* form on the top of the paint but on the bottom.

PAINTING NEW WOOD

If you are following a PPM plan (or at least your own version of one) you may be ready to start painting. The following assumes that all materials and tools have been assembled; the surface to be painted has been sanded; all screw heads have been plugged, sealed, and sanded off; and all seams have been caulked and sealed, and are dry and set.

First primer coat. Apply this coat quickly with a large brush. This type of paint will dry very quickly so work fast and be less concerned with brush marks than with covering the wood completely. It will seal and fill the pores of the wood and provide a good "tooth" for the following finish coats to bite into and hold. You will also find that it will, as it should, raise the nap of the wood and hold it. A final light sanding to remove the nap will give you the super slick finish that you are working toward. If too much of the primer coat comes off with the sanding, give it a second coat of primer. Check the instructions on the can. A quick rub with BRONZE or STAINLESS STEEL wool will smooth it nicely. Never, never, use ordinary steel wool anywhere around a boat. It sticks in the wood and will rust and bleed through the paint. With the primer coat on, fill all cracks and gouges with trowel cement and sand off smooth.

Final Coats

You are now ready to lay up the final coats of finish. Because this paint has, as a rule, much more varnish in it to give the hull that super glossy finish, it must be treated with respect. If you are using paint that has been opened before, be sure to mix it well, adding small amounts of thinner until it is the same consistency as new paint. Strain it through an old nylon stocking. Spread the toe end of the stocking over the edges of the paint bucket. Pour in the paint. Gather the end of the stocking, tie a knot in the open end, and hang it up to drain into the bucket while you do something else. If you are using new paint, be sure to have it well shaken and mixed in the can where you buy it. Pour off into the paint bucket only enough to cover the first third of the brush. Put the cover back on the can, lightly sealed to keep the air out of the can.

Brushing techniques. To load the brush, dip the lower third into the paint. Gently *tap* it on the side of the bucket to knock off the excess paint. Brush it onto the wood initially in any direction to spread it out. Then smooth the paint out in long strokes, all in one direction. If you are right handed you would be wise to start painting the hull on the port stern quarter and work your way around the boat clockwise (if left handed—just the opposite). As you paint, watch for drips, sags, and lamb's tails which indicate you are a little heavy-handed with the paint. Check your work from time to time and stir the paint periodically. If at any time the brush seems to drag or pull, the paint may be getting too thick. Stop and add a small amount of the thinner recommended on the label.

If you are laying up bottom paint, the brush direction is less important than covering the surface. The best way to store leftover bottom paint is to spread it on the bottom of the boat. If you can arrange it, have a helper work ahead of you, cutting in the anti-fouling paint up to the scribing waterline mark. You will find that bottom paint will go on much quicker and cover better when put on with a roller. Use a brush only to touch up the places that the roller can't get.

If your boat has a fiberglass hull, mark the waterline with masking tape first. As soon as the bottom paint is on, remove the masking tape by pulling it off in a downwards direction toward the fresh bottom paint. (This is to prevent the tape from splattering and dripping on the upper portion of the hull.) One drawback with masking tape is that when you remove it, it will lift and pull fresh hull paint off. It is therefore best when you apply the tape to the edge of the paint stop line, to stick only the bottom edge of the tape, leaving the rest of it loose.

Painting over previous coats. Sand old coat to a rough finish so that the new paint will have a "tooth." Prime paint first and fill all cracks and gouges with trowel cement and sand smooth. Proceed as with new surface but without the primer coats. If you intend a major color change, two or more coats of the new color may be called for.

Painting bare metal. A successful paint job in this case depends totally on preparation of the surface. It must be clean of any old paint, oxidized metal residues, and rust. Most important of all, the slightest traces of oil, grease, or wax must be expunged. After the metal has been wire-brushed, sanded, and burnished it should be washed in a powerful mixture of detergent (Oakite, for example) and water. Rinse and dry carefully. From this point on, wear gloves so that you won't touch the bare metal with your hands and leave traces of skin oil on the metal. Wash the bare metal with a commercial degreasing agent, some of which can be found in spray cans. Normally the first coat will be a wash primer which serves the same function as a wood primer by etching the metal and providing "tooth" for the following coats of paint. Choose the primer for the type of metal you are painting. If the metal you are treating is to be under salt water and the final coats of paint are anti-fouling, you may have to put on one or more "barrier" coats of insulating paint. The reason is that the anti-fouling paint uses a form of copper as the toxic agent. Since the hull is not copper, it is crucial to avoid any chance of galvanic action between the fouling paint and the metal. Even with barrier coats there is the danger that scratches may penetrate the protective coats of paint and allow a pocket of galvanic corrosion to start.

When the primer coats and/or barrier coats have dried and set (which is quite rapid in most cases) the final coats may be added. Marine paint manufacturers have developed metal protective paints whose application is complex enough to be called a "system." It is important that you read and follow the directions on the containers. In recent years an anti-fouling paint with tin as the toxic agent has been developed as an alternative to the more common copper based paints to avoid a danger of galvanic action.

Painting new aluminum hulls. Almost everything said previously on painting bare metal applies to aluminum. However, new unpainted hulls still have on their surfaces small amounts of the lubricants that were used in milling the metal at the fabrication plant, and special care must be taken to remove it all. Should any sanding or burnishing be required to remove oxidized remains use *only* stainless steel wire brushes, bronze wool, or non-metallic abrasive paper. (Small particles of metallic abrasives might remain stuck in the aluminum and start a galvanic action pocket.) An anti-fouling "system" for aluminum hulls involves *seven* coats of various paints. A primer, barrier, anti-galvanic guard coat, a second barrier coat, a third barrier coat, and two coats of regular anti-fouling paint complete the job.

Aluminum spars, masts, and booms. These should be carefully cleaned then etched with a teak cleaner. They then can be painted with a clear luster finish such as the Woolsey's product called "Clearlux."

Fig. 4.13 While very fussy to use, this varnish produces excellent results.

Outdrives. It is obvious to the most casual observer that the aluminum alloy outdrives on inboard/outboard powered boats take an awful beating. One of the best preventive maintenance procedures (next to frequent lubrication) is maintaining the protective paint. Here you are faced with two alternatives. If the outdrive is to be totally repainted, you might consider removing the outdrive from the boat and taking it to a commercial paint shop specializing in metal painting and preserving. *Specify* the paint you want and how you want the job to be done. Insist on inspecting each step, especially the removal of all old paint and oxides and the initial etching coat. Some shops have production methods of removing old paint that you cannot equal and it might be a good idea to have them do just that part of the job and do the painting yourself. Several companies supply in spray cans outdrive paint that matches the original paint. One outdrive maker, apparently recognizing the problem, has developed protective coatings and processes for his outdrives. These are quite elaborate and should be very effective. An outdrive project of great magnitude can be saved for the off season. While the outdrive is off the boat, it is a good time to have it completely overhauled before you get into the paint job. Outdrives are complex and overhaul requires extensive mechanical knowledge, a number of very special tools, jigs, and fixtures. So, turn the outdrive over to an authorized factory-trained mechanic for a thorough overhaul. You should know, however, that some of the most unlikely people have managed to overhaul their own outdrives guided by no more than the factory maintenance manual and sporadic expert advise. Special tools were rented as needed and what would have taken a trained mechanic a couple of days took two months to complete, working evenings. The project proved to be a good lesson on the inner workings of outdrives. Several new skills were acquired, to say nothing of the immense satisfaction and feeling of accomplishment.

Interior hull painting. Since the cockpit area gets a lot of wear and tear, expect to do a lot of touching up, with frequent repaintings. Some of this work can be alleviated by choosing the tougher epoxy paints which stand up well to rough usage. In addition to the wear, the area is always under the eye of passengers and guests so you'll want it to look it's best. Should you find that some special area is causing repetitive maintenance problems consider covering it with one of the new super tough plastic materials designed just for boat use. As a rule they are bonded to the surface of the area to be covered with a waterproof cement and the only problems are related to cutting and fitting before bonding.

Cabin interiors. While subjected to less wear and generally protected from the ravages of sun, air, and salt interiors do present special problems. The paint colors should be restful to the eye and still perform the essential object of covering and protecting the wood. Keep in mind that the cabin area is often the first place that wood rot gets started. One of the principal villians is the water from melting ice in the ice box. Keep a tube of RTV handy at all times. This instant rubber sealant is great for stopping leaks around windshields and ports where rain water can leak in and start a bad case of rot. If the cabin overhead is covered with fabric you would do well to consider taking it down as it is a great hiding place for rot.

Brightwork varnishing. Varnished brightwork can be one of your greatest pleasures. Nothing seems to make a boat look better. However, some skippers go overboard on varnishing. It gets so that you don't want to go aboard as there is no safe place to sit or step lest you risk scratching some precious varnish job. Limit the brightwork to those areas not subject to hard usage. There are many types of varnish to choose from. One of the least known and most tricky to apply is "chilled" varnish. This product is kept in a bucket of ice while it is being applied and cannot tolerate bubbles. Once on however, it is truly beautiful and durable. The new polyurethane varnishes are easier to use and come with UV resistant or absorbent chemicals to help them stand up to door use. A totally new product has been developed which is a water-based acrylic varnish. In the can it looks like skimmed milk and smells like vanilla cookies. It is very thin and is best applied with a fine sponge or cloth. It dries in less than half an hour and can be recoated immediately. The finished coat looks as if a sheet of perfect glass had been bonded to the surface of the wood.

A truly successful varnishing job is ninety percent surface preparation, nine percent painting technique, and one percent product. If the wood to be varnished is mahogany, keep in mind that it is a soft wood and must be treated gently before the varnish. Do your sanding with every increasing fine grades of paper and by hand—no machines. Wipe painstakingly with a tackcloth. Since the wood is naturally very porous, apply a filler. There are available combination stain-fillers but they should be used with caution as they darken the wood quite a bit. Apply stain somewhat lighter than you want the final color as the varnish tends to make the stain appear darker. Wipe the stain on with a cloth—don't paint it on. Wipe stain off with a second cloth until the wood nears the color you want. If the stain shows up sanding mistakes, touch up and restain. *Do not* stir or shake varnish; it will create bubbles which will transfer from the brush to the wood. Pour off a small amount of varnish into the varnish cup. Load the brush, which is used *only* for varnish, and draw it lightly across the cup wire. Avoid inducing any bubbles into the brush. Lay the varnish onto the wood and spread it rather than paint it. Painting tends to leave brush marks. Because of the laying on requirement you must take special pains to avoid drips, lambs tails (big drips) and other transgressions. Many coats of thinned varnish are infinitely preferred over one or two thick coats. Varnish enthusiasts rub each coat except the last with pumice and oil. If a dulled gloss is preferred the last coat may be rubbed with oil and pumice also.

Decks. Few small craft skippers are blessed with teak decks although teak-faced marine plywood may yet place it within the reach of all. In any event, deck painting requires a few important considerations. For a heavy traffic area, a tougher than ordinary paint is required. Any deck space on the bow around the anchoring area needs special treatment for that area will get a heavy battering from anchors and chain. Fiberglass boat builders rarely consider precarious footing on wet and slippery gel-coated decks. If slippery surfaces are a problem anywhere on the boat, consider adding non-skid preparations to the paint. (These can be found in any marine paint store.) A second solution is the application of non-skid plastic pads which come with an adhesive backing.

Bilge and below-decks painting. When choosing a paint for one of these areas consider what it will be subjected to—oil drips, grease, gasoline or fuel drips and all sorts of sludge. The selected paint should resist any assault from petroleum products and the answer is a good quality of epoxy type paint. Sweep and dust first. Wash with a strong detergent. Clean out the limber holes that provide drainage between frames. Wait till the painting area is fully dry before painting. Patient drudgery is the best technique. You should not have to do it often.

Painting transoms. If there is any fun involved in small craft painting this might well be the only area. The transom will be assaulted with all of the usual demons as well as exhaust fumes and smoke from the engine. A tough, easy-to-clean paint or varnish is called for in this case. On most small craft the name of the craft (and sometimes its home port) is displayed on the transom. If the object is to paint the name on the boat, and since few of us have either the talent or inclination to do the work free hand, there are several alternatives. One of these is stick-on letters. They come in a variety of colors, sizes, and type styles. If these do not suit you, consider a photographic process that will provide letters in any size and style as photo negatives. These may be cut out and used in a stencil fashion.

5
PLANNED PREVENTIVE MAINTENANCE AND THE WOODEN HULL

As promised, this chapter will offer further details of wooden boat construction. The type of construction determines, to some extent, the methods and techniques required for both preventive and corrective maintenance. However, we are constrained to be brief since less than one-quarter of all small craft manufactured today are of wood. Fiberglass and aluminum are dominating the field due primarily to the lower cost of construction. And good, sound, well-seasoned wood of a type suitable for boat construction is not always easy to find. The cost for such wood is high indeed. When the price of thousands of special fasteners needed to assemble a wood hull is added to this, the costs rise to alarming heights. A man with the skills in hand to build a boat is easily equivalent to a highly skilled cabinet maker and can demand far more money in the field of general carpentry than as a boat builder. Add on the labor costs, and small wonder there is a single wood hull afloat. With all this in mind, if you have a wooden boat needing *corrective* maintenance you had better consider doing it yourself unless you want to undertake the difficult task of trying to find a qualified technician who *can* do the work and won't demand an arm and a leg in payment.

Since the outside of the hull is subjected to the greatest wear and tear, most of both preventive and corrective maintenance centers on the hull area. The vast majority of small craft with wood hulls are constructed in one of the three major methods. These are lapstraked, carvel planked, and plywood planked. (Somewhat more rare forms are the strip planked and the double diagonal planked boats.)

84 Planned preventive maintenance and the wooden hull

Fig. 5.1 Strakes (planks) are lapped one over the other to form a seamless tight fit in this example of lapstraked construction.

Fig. 5.2 Note small frames (ribs) in this inside view of lapstrake construction.

1. THE LAPSTRAKED (OR CLINKER) HULL

This type of small craft construction is an ancient and tried and true tradition. In this mode of construction, planks (or strakes) are fastened to the boat, starting at the keel. Each plank is lapped over the succeeding plank and tightly fastened. Normally, caulking or sealing of the lapped joints is not done as the tight fastenings and natural swelling of the wood after the boat is launched makes for a dry, light, and fast hull. The lapstrake hull has an added advantage in that the laps of the planks tend to reduce rolling from side to side in choppy water. This type of hull, like all wooden hulls, "works" when afloat—that is, the hull members give and take with the pressures of wind and wave. In addition, the fasteners are in a "shear" configuration, which means that the planks put a scissorlike pressure on the shank of the fasteners, and they are under a great amount of stress as a result. Yet even today, fully fiberglassed boats retain the lapstraked design of the hull as if there were lapped planks under the glass—first, to take advantage of the anti-rolling tendency of lapstraked construction, and second, to maintain the traditional and handsome look of the lapstraked hull. Other than severe damage from a collision, most of the corrective maintenance for lapstraked boats will come from seam leaks at the overlaps and are caused by failure of one or more fasteners. Extra problems are sometimes generated by dirt which works its way into the seams and makes resealing difficult. (Methods of refastening a lapstrake will be covered in the next chapter.)

2. THE CARVEL PLANKED HULL

The carvel planked boat may well be the most common form of "skin" construction. Here, the planks are simply butted one to the other and shaped in a fashion that permits a caulking vee between each plank. A caulking, usually of cotton, is hammered into the vee with a special tool, then covered with a seam compound. Due to the normal working of a boat when in the water and the drying and shrinking of the planks when out of the water, these seams tend to open up and leak. Most of the preventive and corrective maintenance of carvel planked small craft involves seam problems, caulking and sealing, cracked and dry-rotted planks, and the usual failed fastenings. Fortunately, recent developments in marine chemistry provide elastic seam compounds which bond the wood and work right along with

Fig. 5.3 Carvel planked hull. Seams are filled with caulking cotton, sealed with seam compound, and then painted.

Fig. 5.4 Carvel planked hull, wet and swollen, shows caulking slightly extruding from seams.

the boat. These compounds set or "cure" from the moisture in the air and, apparently, when immersed in water, only cure better.

3. THE PLYWOOD PLANKED HULL

A substantial number of production small craft have been constructed using one piece of molded *plywood* as a building method. This keeps seam problems to a minimum and makes for a light, strong, relatively maintenance-free, hull. Many of the boat building techniques using plywood were developed under the pressure of war in the 40's when bombers and fighters were made of plywood. Since that time many new and more effective bonding agents have been developed and today's plywood hull is relatively easy to maintain. No matter what you are told, exterior grade plywood *is not* suitable for marine application. It has internal voids or pockets and does not have the integrity or rot-resisting features of genuine marine plywood. If your boat has a plywood hull, most of your maintenance will be preventive paint care with possible repairs for hulls punched by collision. Suggested methods of repair will be covered in the next chapter.

FASTENING METHODS AND MATERIALS

In the case of fasteners, the material used is more important to the average small craft owner than the method. With fasteners designed for marine application, factors to be considered are the holding power, durability in the hostile marine atmosphere, and resistance to rust, corrosion, and galvanic action. At the top of the list are certain kinds (but not all) of stainless steel and monel fasteners, followed by the silicon bronzes, and the copper alloys in screws and boat nails. At the bottom of the list are the galvanized iron screws and nails. To be avoided *at all costs* are brass fasteners, with one exception. Brass is soft metal with little holding power. One of the metals in the alloy brass is zinc which can leach out of the alloy through corrosion. The resulting galvanic action leaves behind a poor piece of junk to do the holding. However, chrome-plated brass screws find their best (and only) use in attaching to the boat various fixtures that are not likely to be under any strain—lighting fixtures and flagpole holders, for example. Deck fixtures such as cleats and chocks which are subjected to heavy stresses should be attached with stainless steel bolts, nuts, and lock washers through a backing plate underneath to distribute the stress. As a general rule, do not economize on the selection and use of marine fasteners, for although the initial tariff is high, they will more than return the investment in the long run.

Fig. 5.5 Example of clinching a nail. The nail is bent over a spike or punch, then driven back into the plank, while holding backing iron to the outside of plank.

Fig. 5.6 After the boat nail is driven through the plank, it is cut and a burr is driven onto the shank, the end of the nail being peened over the burr.

Fasteners are used in several ways. Boat nails are driven into the wood and held by the screw-like protusions around the nail. Other boat nails are driven through the wood and the protruding end of the nail is bent over a punch or nail set and set or clinched back into the wood. The best of the boat nails is anchored, using what appears to be a washer but is called a "rove." The nail is driven into the wood and protrudes through the wood. The rove is driven onto the protruding nail end with a roving tool. The nail is then cut off close to the rove and peened over the rove with many light taps of a ball peen hammer as an assistant holds a

Fastening methods a

Fig. 5.7 This boat wood screw, counter bored and counter sunk, is ready to be plugged with a bung.

Fig. 5.8 Rust from the fastener underneath is forcing the bung out of the counter bore.

backing iron to the nail head—an effective but tedious operation. A number of lapstraked boats are fastened along the length of the plank, where the plank laps over, with bronze machine screws, nuts, and washers. A hole is drilled into the plank every few inches. A bolt or machine screw is inserted into the hole. A nut, over a washer, is drawn up tight with a wrench. The excess end of the bolt is cut off and the end is peened over with a ball peen hammer to prevent the nut from backing off. Planks are also fastened to the frames by the same method thus assuring a tight dry boat. This last method is rare and clinched nailing is more common.

Almost all other fastening methods in wood boats is limited to the common wood screw (marine type), which is driven into a

...nd counter sunk. The hole in the planking ...ally, the screw head is covered with a pro-...e sort and a plug or bung is driven in to ...or both protection and appearance. (If the ...ils are rarely plugged.) After plugging, the ... flush to the planking with a razor-sharp ... flush, primed, and painted. In this method of planking and plank fastening, a lot of preventive and corrective maintenance will involve the removal of protruding bungs (which usually indicate trouble with the fastener behind the bung) and then the removal and replacement of the corroded fastener.

WOOD USED IN PLANKING

Most planking material is cedar, fir, mahogany, and sometimes pine. Choice of wood for planking depends on such factors as strength, rot resistance, and lightness. Spruce is also used but is not at all rot resistant. Although Sitka spruce from Alaska is used as masts and spars, it has for the most part been replaced by aluminum which requires little or no maintenance.

For frames or ribs, the choice is almost exclusively white oak. This wood is heavy and strong. Oak heart wood is very resistant to decay and rot and does not shrink or swell as much as other woods. It holds fasteners extremely well and has superior shock-resisting ability. It is ideal for shaping frames because it can be easily bent using steam or boiling water.

In any event, if your problem is corrective maintenance involving the repair of a wood member of the boat, try to remove the damaged section as carefully as possible so that it can be used as a template pattern for cutting the new piece. Match the new piece with the wood to be replaced where possible. If you can't find the exact replacement, try to go upwards in quality. In the chapter that follows, we suggest several methods of making up a new framing piece using laminating techniques that enable you to produce a piece superior to the member being replaced.

SEAMS AND SEALING

As previously mentioned, it is entirely normal for a wood hull to "work" as it moves through the water. This concept is not restricted to wood hulls alone. Even giant ocean liners are designed to have a certain amount of flex in the hull, for, if they were totally rigid, the movement of the ocean (or any body of water) would soon make short work of them. It is this working, however, that places great stress on all seams that are exposed to water. The seams must be watertight yet work right along with the planking. Formerly this was accomplished by forcing a material called oakum into the seams with a chisel-like caulking iron and a wood mallet. The seam was then sealed with pitch or tar. Later the oakum was supplanted by cotton and various other sealants. New sealers (elastomers) fed out of a tube from a caulking gun as a thick gooey paste sets and hardens into a rubbery substance that sticks to the wood of the seam and flexes right along with the wood. Since this material chemically sets up using the moisture in the air, it gets better after the boat is caulked and launched. When the boat is hauled for the off-season and begins to dry and shrink, the rubbery sealing compound stretches to keep the seam tight.

CHECKING PLANKING—WHAT TO LOOK FOR

When the boat is up on land in its cradle or trailer inspect it carefully for small problems that must be taken care of now to prevent either trouble at sea or a major repair job of corrective maintenance. If the boat is fastened with screws covered and plugged with bungs, look for protruding or loose bungs—this means trouble with the fastener. Cracks in the paint may also indicate a loose fastener. Look for cracks or rot in the planks everywhere but mostly where the planks join. Test the garboards thoroughly with knife point or ice pick to detect soft spots,

Fig. 5.9 Rot city, the wood hull owner's nightmare, shows knife being easily pushed to the hilt in soft punky plank.

Fig. 5.10 Plank separation from the transom is due to excessive caulking instead of needed repairs to leaks.

opened seams, suspicious bubbles in the paint, and the like. A rust streak coming out of the paint will indicate that a fastener has been exposed to the ravages of water and air. If not replaced it will soon give way. Carefully examine the ends of planks where they join the stem and transoms. Butt joints where two planks are joined end to end are always suspect, for not everybody will join them as well as you will. Inside the hull poke around all frames and stringers looking and smelling for signs of wood rot, bad cracks, streaks of rust, and corroded fasteners.

BILGES AND THROUGH-HULL FITTINGS

We have mentioned the basic problems associated with bilge maintenance. Because the bilge is essentially out of sight, it is often out of mind. A minor problem, not caught in the early stages, can quickly grow into a major corrective maintenance project. One easy problem to cure is a plugged limber hole. These are small holes drilled or cut in the frames at the bottom of the bilge so that drainage can take place from the bow down to the stern. The limber holes can easily become plugged with dirt, and the pool of stagnant fresh water which forms is an open invitation to the spores of rot. As a preventive measure, install a light bronze or galvanized chain that runs continuously through each limber hole from bow to stern. The chain is left threaded through the limber holes at all times. A few yanks on the chain, and the holes are cleaned out. While checking the bilges, be sure to cast a hard eye on all through-the-hull fittings. Pump suction and discharges, drains, depth sounder heads—anything that goes through the hull—should be inspected for watertight integrity and to make sure no galvanic action or corrosion is taking place.

DECKING

Decks, especially those exposed to weather, take a beating. Sun, rain water, foot traffic, and the general abuse by anchors, chains, and other heavy objects, will keep you busy with preventive maintenance. Most of the problems with decks, as we have said before, come from small leaks that are difficult to find and plug. They cause those annoying small drips that make the inside of the cabin damp, moldy, and uncomfortable in wet weather. Indeed, a tight dry deck is a thing of beauty and a joy forever.

6
CORRECTIVE MAINTENANCE AND THE WOOD HULL — Repairs Required Due to Neglect and Damage

When in spite of all your efforts in PPM there comes a time that some corrective action must be taken in order to return the boat to a safe and seaworthy condition, some of the following will be helpful. It is not possible to cover every contingency in this limited text, and because of this we have selected procedures that embody as much method and technique as possible and that will be applicable to many repairs within the capability of the average skipper. Expect that there are some jobs you will have to turn over to the boat yard or a professional, either in whole or in part.

Repairs for wood boats are subject to *one* fundamental problem. There are few, if any, simple straight lines anywhere on a boat! What this means to you is that to cut and fit a new piece of wood for a boat is rarely a matter of drawing a couple of straight lines on a bit of wood and cutting away everything that is not the part you want. It is essential that you clearly understand this, and when you do, most of the problems of wood boat repair will be more a matter of patience than extreme degrees of skill and craftsmanship. Yachtsmen are fond of saying that the worth of a good boat can be easily evaluated by observing the quality of the "joinery." Joinery depends upon the degree of skill and craftsmanship of the builder (repairman). Wherever one piece of wood on a boat is *joined* to another, the method used and the degree of skill required determine the quality of the "joinery." The master boat builder, in repairing a wood boat, will measure, cut, and fit the needed new piece of wood using techniques and skills he has acquired through years of practice. Two of the most important skills are "spiling" and "templating."

Perhaps you have watched someone cutting out a dress from a paper tissue pattern. The pattern is, in effect, a template. In dressmaking, the template is already prepared and sized and needs only to be pinned (with great care) to the cloth for cutting. In our case we have to make the template *before* we can use it to transfer the outline to the new wood prior to cutting. Spiling has much the same objective—that is, the transfer of certain dimensions within the boat to a pattern, and the subsequent retransfer of those measurements from the pattern to new wood. The professionals use either a scribing compass or a pencil compass to transfer the measurements to the pattern. The precise, lucid, and clear explanation of how this is done is beyond the scope of this chapter. Suffice to say that years of experience as a master boat builder are required. If, however, you have little or no experience, substitute one or the other of the following methods. Rather than attempt to explain the method you should consider, we have illustrated it through a series of photographs.

The vehicle for this demonstration is a "Turnabout" sailboat of which several thousand were built not long ago on the North Shore of Massachusetts. The boat shown suffered from a terminal case of dry rot and was, in the opinion of professionals, "beyond redemption." The entire transom together with the supporting frames had to be replaced. We began by removing all the paint from the bottom and insides to uncover any further evidence of rot. The plywood skin was found to be solid but had suffered minor damage from previous and futile attempts at repair. All wood screws and boat nails were carefully extracted. Each piece of wood, no matter how badly rotted, was carefully removed and set aside for use as a pattern to be traced on new wood so that wherever possible, the need to use spiling techniques could be avoided. In several instances what appeared to be a perfectly sound piece of wood turned out to be spongy and useless when the many layers of paint were removed. Rot had started in the transom frames and had spread to the plywood transom itself. From this evidence, the oak keel became suspect, and when it was removed rot was confirmed. A piece of white oak was obtained at a local lumber yard and ripped to size on a table saw. The otherwise difficult bevel cuts were nothing to the table saw which reproduced them with ease. The slot for the center board was cut using a sabre saw and finished with a heavy wood rasp. The new keel was set in one of the new curing compounds, as flexible rubber cannot be separated from the wood

96 Corrective maintenance and the wood hull

Fig. 6.1 As evidenced here, rot has no respect for size. The transom from this small sailboat is found to be nearly rotted through underneath the transom braces.

Fig. 6.2 Transom braces from sailboat rotted through and held together by paint.

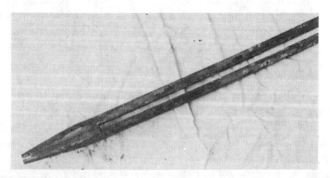

Fig. 6.3 Keel from same boat suffering from rot.

Corrective maintenance and the wooden hull 97

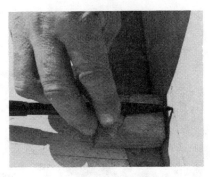

Fig. 6.4 How "spiling" block is used to make spiling batten.

Fig. 6.5 Adjustable "T" bevel used to take bevel from old frames and transom edges as compared to regular square.

without damaging the wood itself. Screw holes were counterbored and silicon bronze wood screws were sent home with a bit brace in which a screwdriver blade had been chucked.

With the new keel installed and the boat tightened up, the old transom no longer fit precisely. The spiling technique was used to make a better fitting pattern for the new transom. A piece of scrap thin plywood was rough cut to fit the transom space and tacked in place. A spiling block (2" × 2" scrap of oak) was used as shown to transfer the exact outline of the new transom to the thin plywood. When the pattern was complete it was removed from the boat and tacked down to a piece of scrap plywood to see how well the pattern would fit before any expensive marine plywood was cut. The spiling block was used to transfer the outline from the pattern onto the test stock and the

Fig. 6.6 Test transom fits well first time out.

cutting out was done with a sabre saw. All saw cuts were made following a sound professional practice of always cutting on the *waste* side of the pencil line. Wood in excess can easily be removed but putting it back on is another thing entirely. An adjustable "T" bevel was used to measure and take off the bevel from *all* edges of the old transom. A small block plane was used to cut the bevels a little at a time — fitting the test transom to the boat after each few cuts. Slowly and surely (taking far more time than the professional would take) a closer and closer fit was achieved in the test stock. Soon it was obvious that the spiled pattern would, if patiently followed, produce a perfectly fitted new transom. New frame pieces were cut from fresh white oak stock (flooring scraps from the lumber yard) using the old frames as templates. The versatile sabre saw made short work of the otherwise difficult cuts and all new wood was thoroughly soaked in wood preserving Cuprinol. The new frames were bonded to the transom using an epoxy marine glue and silicon bronze screws. The transom assembly, bedded in the same elastomer as the keel, was installed and the skin of the boat screwed to the oak frames of the transom.

In the preceeding case history the repair was effected with a small amount of spiling and the careful removal of the rotted frames so that they could be used as patterns. Had there *not* been a piece of scrap plywood handy a large section of cardboard could have been used. It could have been stiffened with wood lath and tacked in place for use as the spiling batten. This would save trying to find the right size of scrap.

Corrective maintenance and the wooden hull

Fig. 6.7 "Dave's Weird Wonder" will faithfully reproduce any complex curves and shapes.

If the spiling process seems too difficult, you might try making up one of our "Weird Wonder Spilers." This is nothing more than a section of cardboard cut to expose the holes in each side. Small dowels are inserted lengthwise in the holes. The cardboard is tacked or stapled close to the complex surface to be spiled and the dowels are pushed through the cardboard until they just touch the surface to be matched. It is a simple matter to remove the cardboard and again tack it in place on the new stock to be cut. Lines are traced onto the new stock with the ends of the dowels as points.

FASTENING MATERIALS AND METHODS

Once the new piece has been cut, fitted, preserved, and set in place, it must be fastened. There is a wide choice of fasteners and methods. Several of those that are tried and proven will be shown. For quick and temporary repairs in an emergency, plain nailing will get you safely to a place where a more permanent repair can be effected. However, nailing must not be sneered at, as it is an old acceptable method of fastening a boat together. Unlike house nails, boat nails (usually) are made of copper.

Boat Nails

As illustrated in Figure 6.8, nails come in a variety of sizes and styles. The nail with the threads similar to a wood screw should be used when the opposite side of the wood is inaccessible. These nails hold well if the wood that they are holding does not work excessively. If you choose to use this type of nail for any reason be sure to drill a small diameter pilot hole first. Rub the shank of the nail on a bit of softened soap or beeswax and drive it home. The pilot hole will help to prevent the splitting of the wood and the bending of the soft nail. As a rule these nails are countersunk only by a final hammer blow, which is to set the head just below the surface of the wood so it can be covered with putty or paint.

The next and somewhat more effective method (as far as holding is concerned) is "clenching" or clinching the nail point after it has been driven through the wood to be fastened. Unfortunately, this method requires two people, as do several of the better methods. Similar to the first method, the lubricated copper nail is driven into the wood through a pre-drilled pilot hole. A helper holds a backing iron (which is nothing more than a large hammer or piece of heavy iron) against the head while the pointed end is bent over a nail punch or spike, and drives the point of the nail back into the wood. The nail is then said to be clinched into the wood and will hold effectively. Keep in mind that a clinched nail is difficult to remove, as you may find out in an older lapstraked wood hull.

The third method uses the same nail but adds a copper "rove" which is similar to a copper washer as shown. When the nail has been driven through the wood, the rove is forced onto the protruding nail. (The nail must be driven since the hole is a smaller diameter.) The rove bites into the shank of the nail and grips it tightly. The point of the nail is then snipped off using

Fastening materials and methods 101

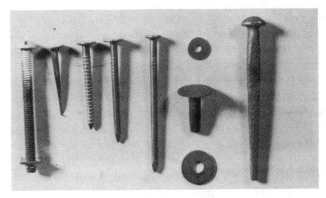

Fig. 6.8 A variety of boat nails. Note the old fashioned square cut boat nail.

Fig. 6.9 Boat nail fixed with burr and peened over.

heavy cutters. A few light taps with a ball peen hammer rounds off and sets the nail tightly against the rove. Again, it is necessary to have an assistant holding a backing iron against the head of the nail during the setting of the rove to the nail and the peening over of the cut nail. Setting the rove onto the nail requires a special tool not often readily available but you can create one rather easily by cutting the threaded end of a hefty iron stove bolt off with a hacksaw. Drill a hole into the center of the cut face of the bolt. Make the hole slightly larger than the diameter of the nail on which the rove is to be set. It works — and it's cheap.

Nuts and Bolts

Nut and bolt combinations can also be used, primarily in lapstraked boats, and have an added advantage in that they can be retightened when necessary. As always, pre-drill a pilot hole

102   Corrective maintenance and the wood hull

slightly smaller than the diameter of the bolt. The bolt, usually a slotted flat head screw, is driven into the pilot hole and threads itself into the wood. A helper on the inside of the boat threads a nut and washer onto the bolt end while the bolt is tightened from the outside. When this has been set up tightly (enough so that the washer bites slightly into the wood), the bolt is cut off near the nut with heavy cutters and peened over so that it cannot loosen and back off. Both the nail rove and the bolt/washer can be removed only by grinding or drilling free the nut and rove. Small grinding stones that can be chucked into an electric drill are now available along with a variety of other accessories that make the removal of these two fasteners quite easy.

Rivets

These popular fasteners are used extensively on aluminum hulls and fortunately for us a new tool is available called the "pop

Fig. 6.10   "Pop" rivet tool with wide variety of steel and aluminum rivets.

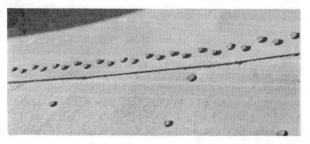

Fig. 6.11   Thin skinned aluminum hulls are bonded together with aircraft type rivets as shown here.

riveting gun," which makes "blind" riveting possible. That is, rivets can be set into the hull without the necessity of using a backing iron or upsetting tool on the inside of the boat. Rivets come in a variety of sizes and applications and, best of all, most of them are made from an aluminum alloy so they can be used on an aluminum hull without causing any galvanic action. Some rivets are hollow, and once set into the boat they must be filled to prevent leaks.

Screws

By far the most popular fastener on wood hulls is the wood screw. However, not just any iron, steel, or brass screw will do. Iron or steel screws, galvanized or not, often join in galvanic action with any other nearby metal. Also, they rust and corrode. Brass is soft and has poor holding power, and the zinc in the alloy rapidly leaches out, leaving behind a soft powdery metal of no use whatsoever. The only screws with any holding power that are impervious to galvanic action and corrosion are those whose alloy is silicon bronze, monel, or some form of stainless steel. Expect to pay a premium price for these fasteners. However, their value in the long run cannot be disputed. Not all stainless steel alloys are suitable for marine use, but this is not to say that the best grade of galvanized steel screws cannot be used. As a matter of fact they are still one of the most popular fasteners among many builders of wood hulls.

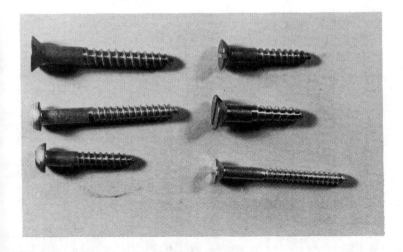

Fig. 6.12 The best wood screws are silicon bronze.

104　Corrective maintenance and the wood hull

Since the removal and replacement of rusted galvanized wood screws is pure preventive maintenance, the sequence for this activity is shown in the accompanying photographs. The first evidence of trouble is a rust streak which immediately reappears even if painted over. Soon the bung or plug swells and loosens to drop away. If the problem is not attended to, soon the head will rust away and it will be nearly impossible to back out the old screw and replace it. You may be able to drill out the broken shank. A better method is to drill a new hole immediately above the offending screw. A heavy punch is used to bend it over as in clinching a nail. The bent screw is given a dose of epoxy glue and a new screw is set under or over the old one. Both holes are then bunged with plugs carefully set in epoxy glue, cut flush with a chisel, sanded and painted.

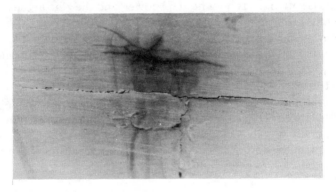

Fig. 6.13 "Trouble a-brewin." Due to rusting screw underneath, bung will soon be forced out of plug hole.

Fig. 6.14 Bung marked for removal during inspection will hasten removal method.

Fig. 6.15 Driving screw into bung forces it out.

During your inspection prior to formulating the preventive maintenance plan, look for any suspicion of rust streaks bleeding through the paint on the hull and check for loose or swollen bungs. Each location should be marked with a bright colored crayon or chalk and replacement calculated into your plan. Begin by removing loose bungs with a sharp pointed tool. If still tight and holding, drill a pilot hole into the bung till the tip of the drill bit just touches the screw. Screw in a wood screw which will force out the bung when the point engages the old screw. With a sharp pointed tool such as an awl or a prick punch carefully scratch away the rust from the screwdriver slot. To break the seal between the threads of the screw and the wood, set a heavy screwdriver whose blade just fills the slot and give the screw one or two smart raps with a hammer. Chuck a proper sized screwdriver bit in your bit brace and while applying the most pressure possible, rap the brace handle with the heel of your hand to start out the old screw. With a bit of luck and a lot of patience most of the old rusted screws will come out with this method. Replace the old screw with one of better quality or replace it with the same type but slightly longer screw to make up for the loss of holding power. Re-bung with a plug set in epoxy glue. When setting the bung, keep the grain of the bung in the same direction as the grain of the plank. Unless you are doing an extensive plug and screw replacement job do not buy a plug cutter for your drill. Most marine hardware stores sell small packages of pre-cut bungs in the size that you need.

GLUES—ADHESIVES, BONDING AGENTS

The PPM skipper has a wide choice of easily applied marine glues (or, if you insist, adhesives). For every maintenance task, whether it be corrective or preventive, there is a glue that will do the job. Often the glued, bonded joint, or repaired part will be as good as the material it bonds and sometimes better. Any glue sticks because it (1) hardens and (2) sets or cures. This is done by cooling to evaporate or "polymerize" the solvent. Animal glues which are melted before use and harden after cooling are useless around boats. Rubber cement evaporates its solvent and hardens but its use is limited on a boat. The epoxies polymerize when a separate chemical catalyst "prods" the other chemicals into forming chains of molecules which bond to everything they touch including themselves. They "cure" after setting by internal chemistry, by using the moisture in the air or heat.

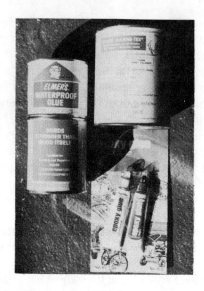

Fig. 6.16 Examples of waterproof bonding agents.

The advantages to epoxies are many—they don't shrink after setting, can be sanded and painted, and are harder and stronger than the materials they bond. They are electrical insulators, they come in a variety of colors, and are usually guaranteed not to wear, tear, leak, or rust.

When a bonded joint fails, it is usually the result of the wrong glue. For example, a glue that uses a solvent might attack the material to be bonded. In some glues there is an acid base that might attack the material and in other cases the acid in the wood (oak, for example) might attack the glue. Some glues may stain the wood. A very porous wood might absorb the glue leaving nothing behind to make the bond. For bonding metal to metal extra care must be taken to prepare the surface to provide "bite" or "tooth" for the glue to adhere to.

Glue may be brushed, sprayed, rolled, dabbed, layed on in sheets, put on from a heat gun, or squeezed out of a tube in a caulking gun. Generally, when one of the epoxies is applied with a brush, you can expect to have to throw the brush away. With any glue there are three general requirements. First, the joint must be clean—really clean. Second, the glue must be chosen for the job it must do and the materials it is to bond. Third, it comes with directions which must be read before using.

There are several precautions you must observe. Some of the new glues have a limited "shelf life"—that is, once they are manufactured they must be used by a certain date. Others, of the two part type where a catalyst is mixed with the body of the glue, have a limited "pot life," which means once the catalyst has been added they start to set and harden and there is a limited amount of time to spread, join, and clamp the material. Some set up and harden quickly and are said to "kickover" in the container. All of them *may* contain chemicals that irritate the skin, are dangerous to the eyes, or have toxic fumes. If precautions are not observed still others are a fire hazard, either because the fumes are explosive or because of the heat they generate while curing. *Read the directions on the can.*

WOOD HULL REPAIRS

The following specific examples of repairs to a wood hull were selected to illustrate method and technique rather than specific types of repairs. A complete coverage would require an encyclopedic collections of books.

Patching a plywood hull. When all laminates crack through (however unlikely), or an outright hole has been punched in the hull with a resultant de-lamination of the layers of wood, the following method has been found to be very satisfactory. Carefully draw lines at the limits of the damage on the outside of the hull. Whenever possible, mark out in a square. Draw a second set of lines at least two inches beyond these limits so that the repair will be seated in good strong wood. If the damaged section of plywood is heavily curved, a whole new section of plywood must be formed and layed up. Such a repair must be turned over to a professional with the special tools and materials required for the job. For the patch however, carefully check the inside of the boat to locate any ribs, stringers, or risers that might be in the way as you cut out the damaged section. Drill a pilot hole large enough to admit a sabre saw blade or hand keyhole saw. Carefully saw out the damaged section of hull following the lines drawn on the hull. Take your time. Make the cuts as true and straight as possible. Stop frequently and check inside the hull to see if the saw blade is approaching a hull support member that you *do not* want to cut. If that is the case, stop, skip over the area behind which there is a rib and drill a new pilot hole on the other side and start again with the saw. Use a razor sharp chisel to complete the uncut sections of sawn lines. If there are any fasteners left in the damaged section, they must be carefully removed (refer to the previous section on fasteners).

When the damaged hull section has been cut and all fasteners removed, inspect for further damage to any rib sections or stringers beneath. If these are inaccessible from inside the boat—as they often are—they can be repaired from the outside the boat with a little ingenuity (see how to form and lay up a sister frame in a later section). These must be repaired before proceeding with the hull patch.

To complete the patch you have three alternatives. The first and most difficult involves cutting a scarph bevel at the cut-out edges of the hull. A matching scarph bevel must be cut in the patch. It is best to cut the bevel on the patch section first as any mistakes at this point can easily be corrected. The hull scarph is rough cut with a fine toothed saw and finished using a rubber backed sanding disc chucked into an electric drill. The scarph bevel should be not less than three to five times the thickness of the plywood. While cutting, stop frequently to trial fit the pre-cut patch. Stop just before you achieve a perfect flush fit. Mix up enough marine glue to cover the scarph surfaces well. Set the patch into the hole and apply bonding pressure. (If an epoxy glue is used, pressure will not be required.) If the inside of the patch is accessible you can screw a strong eyebolt into the patch and connect a turnbuckle to the eyebolt with heavy wire (or anything that will not stretch). The turnbuckle is taken up tightly to pull the scarphed patch against the hull. Pressure can also be applied in a number of ways outside the hull. Run a chain around the hull, padding it where it touches the hull. Insert wedges under the chain at the patch and drive up tight. Or, use a brace that will apply enough bonding pressure until the glue has set. Next, sand down the entire outside patch area until it is flush and smooth to the hull. When primed and painted it will have the strength and the appearance of the original unbroken sheet of plywood.

The second method, while quite a bit easier, requires the use of an unsightly backing block to be fixed to the inside of the hull before the final patch is cut and fitted. The backing block can only be used if the area behind the patch is free of any frames or other construction. The backing block is cut from the same thickness of plywood about one inch or more larger than the hole to be patched. If it is possible to work inside the boat, drill pilot holes for wood screws of proper length around the edge of the backing block. Set the backing block in glue around the edges where they contact the inside of the hull. Insert and screw the wood screws tightly into the hull. Using the damaged section that you cut out as a template, mark out a patch on new marine plywood stock of the same thickness. Cut the new patch on the waste side of the pencil marks and fit it to the hole in front of the backing block by sanding or planing gradually to achieve the tightest fit. Generously spread glue on the backing block and the edges of the patch hole and set the finished patch into the hole. Screw the patch tightly to the backing block with marine wood screws. If the plywood is thin, don't countersink the screw heads — just tighten them enough to allow the heads to sink flush with the wood or a little below. If the plywood is three-quarter inch, countersink the pilot holes for the screws, only enough to permit covering the screw heads with trowel cement. Sand, prime, and paint and the hull is as good as new.

The third method still requires some form of a backing block but it need not be thick or unsightly. The entire repair can be completed in the same manner as patching a hole in a fiberglass hull (instructions in the following two chapters). It is relatively easy to accomplish and makes as good a patch as the other method, maybe even a little better.

If the problem is an unsightly mar or scuff that does not deeply penetrate the plywood, consider marking out the area in a diamond shape and cutting out the damaged area only to *one* depth of the ply. Cut a thin strip of wood slightly thicker than the ply to fit the diamond shape, set in epoxy glue and sand flush after the glue has set. Alternately, you can fill the void area of the damage after *all* paint has been removed and fill it with a putty made of sawdust and epoxy glue. Sand it flush after it has set.

Patching a lapstraked hull. Because of the method used to lay up the planks of this type of hull, repairs to the strakes or planks can be a bit of a challenge. The lapstrack hull is of light weight construction and normally has many small frames spaced closely together. Since there is no caulking between the seams (aptly called dry seams) there are many more fasteners used at closer intervals along the edge of the planks where they overlap. If the fasteners are copper nails with roves, the roves may be ground off with a small grinding wheel chucked into an electric drill. Using a heavy punch and hammer, drive out the nail from the inside of the boat. An old knife blade might be forced between the laps and driven along with a hammer, shearing soft copper nails. If the fasteners are wood screws, the heads must be uncovered one by one, the slots cleaned out carefully, and the screws backed out of the wood. With the fasteners out of the way, drill a pilot hole to admit a short bladed sabre saw near the damaged area. Make vertical cuts in the strake several inches either side of the damaged area. Stop saw cuts above and below undamaged strakes. The balance of the cuts can be finished off with a sharp chisel. With the damaged section fully cut and all the fasteners removed or cut, the section may still be held in place by old paint and the clamping action of the other strakes and will have to be prized out. If you have the patience to try it, we recommend that you scarph the ends of the cut strakes and cut matching scarphs in the new wood patch. The repaired strake should equal a new plank in strength and appearance. When you have achieved as close a match as you can, proceed as with the scarphed patch in the plywood hull.

If the damage is confined to a single strake and is a puncture fully or partly through the strake leaving the strake plank otherwise sound, an easier patch method is to cut out the damaged area using one of the new hole cutting saws chucked into an electric drill. This will cut a neat, perfectly round plug out of the damaged strake. The edges of the cut hole may then be beveled (or scarphed again) with a wood rasp, filing with the larger area on the outside of the boat. Cut a wood plug from new slightly thicker stock using a sabre saw and bevel to match the hole in the plank. The plug should match the hole closely and leave a small amount of protruding wood. Lather the edges with a good epoxy marine glue, and set in place. After the glue has set, sand the plug flush. You can then proceed with priming and painting.

Most of the problems with lapstrakes relate to leaking and weeping seams. The relatively thin planks and light frames are designed to work and give. Often this causes the fasteners to loosen along the seams. No amount of fastener tightening seems to help. As the seams work with the movement of the boat, dirt and other foreign particles work into the lap seams to prevent a perfect wood-to-wood seal. No amount of caulking seems to work even with the new two-part elastomers. (There is no caulking space between the seams and the material will not adhere to anything but clean wood.) To remove dirt in the seams, the

Wood hull repairs 111

fasteners must be removed for several inches either side of the leak. Pry open the seam with sharp pointed flat wood wedges, creating space enough to insert an old hacksaw blade to clean the length of the seam. Insert new fasteners—in new holes if possible—and draw up tight. Fill old fastener holes with whittled wood plugs daubed with epoxy glue, cut off near flush, sand, and finish.

When the seam leaks extensively for some distance along its length, you may have to resort to a trick the professionals use. You can do most of the work yourself but the initial step may have to be done by an expert unless you have access to an electric powered router. The bottom edge of the strake over the leaking seam must be changed from a softly rounded edge to a sharp

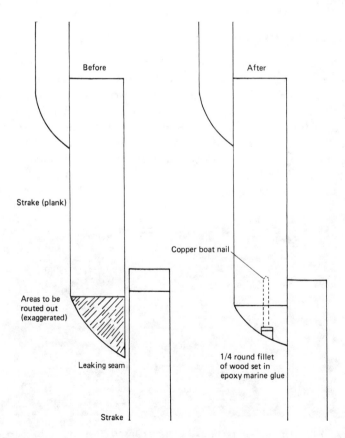

Fig. 6.17 Sketch of method to seal lapstrake seam.

90 degree edge as shown in the Figure 6.17. The interior angle formed by the junction of the two overlapping planks must be carefully cleaned of all paint and primer down to bare wood. A long piece of quarter round wood is then set into the junction of the strakes using epoxy glue or it may be nailed to the top plank with copper boat nails. The ends of the fillet (as it is called) are smoothed to soft curves and sanded flush, primed and painted. If you do not have a router but believe that you can manage one, rent one from a tool rental shop. Be sure to get the right cutting bit to go with it—otherwise you'll either have to call in a helpful friend or hire a professional to make the cuts on the boat. The rest you can do yourself. When the fillet has been bedded in epoxy glue, sanded and painted, it will be a nearly invisible non-leaking seam.

CARVEL PLANKED HULL REPAIRS

Much of the preventive maintenance required with this type of hull construction involves leaking seams and rusting fasteners. This type of hull construction does not work as much as the lapstrake and many of the problems seem to come when the boat is hauled for the season and the planking dries and shrinks. The caulking becomes loosened and may be cast out of the seam. Before launching for the new season some light recaulking may be required, and after the boat is launched it should be allowed to soak for a few days "on the pump." The boat may leak badly

Fig. 6.18 Drying and shrinking has caused this seam to cast its caulking.

at first but gradually, as the planks of the hull absorb water and swell, the seams tighten and the hull may be said to be dry. This swelling and shrinking of the planks place a severe strain on the caulking. Some skippers cut up old rugs and put them in the bottom of the boat when it is out of the water. Soaking the rugs from time to time with salt water will keep the seams tight. Of course, this method requires periodic trips to the boat to keep the rugs wet, but there is much to be said for wet storage in the case of carvel planked boats.

Cracked Planking

To repair a cracked plank where the crack is not extensive, the first step is to arrest the crack by drilling a hole at each end. This technique is also effective in fiberglass and aluminum hulls. Scrape or otherwise remove the paint down to the bare wood and force epoxy glue into the crack. When the glue has dried and set, sand flush and paint. The drill holes can be plugged with whittled plugs and glued. Do not *pound* the plugs into the wood lest the crack be restarted.

If the crack runs vertically across the grain of the plank, consider the use of a butt block inside the boat over the cracked area. Assuming the cracked area of the plank is accessible from the inside of the boat, carefully scrape and sand the inside of the plank down to bare wood. Wash the cleaned area with a good degreasing thinner such as acetone. Cut and fit a butt block well buttered with epoxy marine glue over the inside cracked area. While the glue is drying and setting, countersink and counterbore wood screw pilot holes from the outside into the butt block and screw in at least five of the best quality marine wood screws you can afford. Seal and plug the screw head holes with bungs and epoxy glue. If the crack is extensive with loss of wood around the cracked area you are facing the *removal* of the broken plank section and replacement with new wood. You will seldom have to replace the entire plank. However, if this *is* the case and the plank is curved, you may have to turn to the professionals. Otherwise, you can proceed in the following manner. First, remove the damaged section of plank. Mark off the section to be cut in between frames so that butt blocks can be used to tie the ends of the plank together. Drill a pilot hole on the bottom edge of the plank section to be removed and make a vertical cut with a sabre or keyhole saw. Make sure that the saw blade cuts only the plank and nothing inside the boat. Next, remove the

114   Corrective maintenance and the wood hull

fasteners. One of the quickest and neatest ways to accomplish this is to chuck up a small diameter hole saw in an electric drill and cut a hole right around each fastener. The plank can then be easily tapped out from the inside of the boat. This method also aids in retaining the cut-out section of plank as a template which will be needed to mark out and cut the new plank section. Once the damaged section of the plank has been removed, the fasteners (with their collars of wood) may now be removed using a pair of heavy vice-grip pliers to turn out the old screws. Carefully inspect the frames behind the removed section of plank for other damage. (Repair of damaged frames follows this section.) Make the repairs *before* replacing the damaged plank. If the damaged section of plank was in such poor condition that it cannot be used as a template, then it will be necessary to make a marking template which is a simple but exacting task. Fit a large piece of heavy cardboard into the cut-out section of the planking. Leave a space of one inch around the entire area of the cardboard. Tack or staple cardboard onto the frames of the area where the new plank section is to go. Using a two-inch block of wood and a sharp pointed pencil, slide the block of wood along the edges of the planks as you draw the pencil along with the block, marking out a line on the cardboard. Do this slowly and carefully so that the outline of the area will be transferred to the cardboard. Remove the cardboard and tack or staple it to a piece of new wood the same thickness and type as the old plank section. Before you begin transferring the lines from the template to the new wood, glue a small piece of cardboard to the edge of the block of wood so that as you slide it along the line on the cardboard, to mark out cutting lines on the new wood it will not tilt on the cardboard. Now slowly and carefully slide the block along the marked lines on the cardboard and transfer the lines to the new wood with the pencil. Clamp the new wood on a flat working area and being careful to cut on the *waste* slide of the marking lines, cut out the new plank section with a sabre saw. Trial fit the new rough-cut plank section. It should be slightly oversized all along the horizontal edges. Using a razor-sharp small block plane and great patience, slowly remove small amounts of wood from the horizontal edges and *one* of the vertical edges, stopping frequently to trial fit the new section. When the newly cut section fits snugly, you are ready for the next few steps.

The caulking bevel is next assuming the plank you are repairing is beveled. Take the bevel angle from the old piece using an adjustable tee bevel. As a rule, one-third of the edge is left flat and two-thirds are planed off to make the caulking bevel. Mark the bevel lines both on the edge of the plank and the face of the plank. Again clamp the new stock on a flat surface or in a wood vise and plane off the caulking bevel. Although a large plane will remove the wood faster, the small block plane is less prone to error and easier to control. Reset the new plank section back into the boat skin.

Next, cut butt blocks to fit at the ends of the new plank section inside the boat. Carefully remove all paint down to clean wood on the ends of the old sound planking where the butt blocks will be attached. Leave the cleaned off wood area rough enough to give it "tooth" for the glue. Chuck the combination pilot hole drill, counter sink, and counter boring tool into an electric drill and drill the holes for the new wood screws and bungs in the new plank section. Soap the threads of the screw lightly so that they will drive smoothly with a screwdriver bit chucked into a bit brace. The butt block may now be set inside the boat against the new plank section ends where they join to the old planks. Using epoxy glue between the butt blocks and the planks, set them in place with at least two to four screws inside the boat through the butt block and into the plank ends. Drill pilot holes and counterbores for plugs on the outside of the boat into the plank ends and butt block. Plug all screw holes with bungs set in epoxy glue, cut the bungs flush, sand the new wood, prime it, then complete the caulking. Seal with one of the new

Fig. 6.19 This combination woodscrew pilot drill, countersink and counterbore is an all-in-one operation tool.

elastometric marine sealers and final paint to match the rest of the hull.

The preceding repair should be as sound if not more so than the rest of the planking. This type of carvel planking repair is restricted to relatively short sections of uncurved planking. If several sections of planking close to each other must be replaced be sure to stagger the lengths so that the butt joints do not line up vertically. When the plank to be repaired must follow a radical and rounded curve of the hull you may have to consider calling in a professional to cut, steam bend, and fit the new section. In other sections of the planking the inner surface of the plank will require planing to a concave shape so that it will fit snug against the curves of the frames. This will also require the aid of a specialist.

REPAIRING DAMAGED FRAMES

The repair of cracked, rotted, or damaged frames (ribs) is normally easier than plank repair. In the great majority of cases the broken or damaged section is left in place (unless rotted) and a "sister frame" is positioned alongside the old frame and screwed or bolted in place with the ends bonded with epoxy glue to the old frames. There are several methods that allow the owner-skipper to do the job adequately. We shall describe a few methods here and you can decide which one suits your problem and the way you wish to solve it.

If the frame is badly cracked or broken, first (mostly for appearance) dope the broken ends with epoxy glue. Clamp the ends between wooden blocks covered with wax paper or plastic so that the clamp blocks will not bond to the damaged frame as the glue sets and cures. If the damaged frame is straight and uncurved, measure and cut a new frame section from tough, well seasoned wood, that is as strong as the damaged frame. The length of the section should overlap the damaged section by four to five inches on each end. Soak the new wood in a good preservative *before* you install it. Either bolt it to the old frame with heavy bolts, washers, and lock nuts, or use wood screws. Either way, use epoxy glue along the entire contacting surface, making sure that the paint, oil, and grease are cleaned from the broken frame before you set in the new frame alongside. Give the finished new wood a final soaking with preservative before painting.

Repairing damaged frames 117

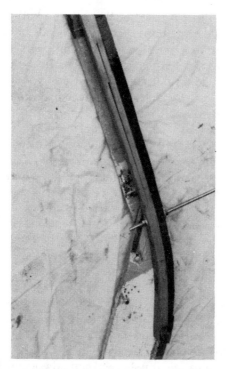

Fig. 6.20 How a "sister" frame might be cut. Lateral cut allows repair frame to bend easily and conform to hull curve as marine glue bonds and sets frame. Threaded stock bolt is removed later.

When the problem is a section of frame in which you have found wood rot, the entire rotted section must be *cut out*. Don't try to save anything with the slightest sign of rot in it because it will infect the surrounding wood. Soak all bare wood in the area with preservative. Cut and fit a sister frame if the old frame was straight. When the frame to be repaired follows the curve of the hull, a different strategy is called for. Both of the following suggestions avoid any necessity for steam bending of frames and will provide a repaired frame that is stronger than the original. By far the easiest method is to cut a section of oak stock that is two inches wide and an inch thick. On a table saw make a lateral cut lengthwise in the one inch dimension, two-thirds the length of the stock. Position the new frame flush to the old frame. Drill bolt holes through the hull and the sister frame. Heavy steel stove bolts, with a wood block under the heads to protect the finish on the outside of the hull, are set through the hull and the sister frame. The number of bolts required depends on the length of the sister frame, but set at least one bolt every foot. You will have to drill from the sister frame to the outside of the boat to line up the bolt holes long enough to insert the bolts. If you cannot find bolts long enough, use threaded rod that can be purchased in nearly all hardware stores. Set the bolts in place with large flat washers at each end. Dope the saw cuts in the sister frame with epoxy marine glue and begin to tighten the nuts and bolts. The saw cuts in the sister frame will allow the new frame to bend without breaking as the bolts are slowly drawn up tight. When the epoxy glue has set and cured, fasten the planks to the sister frame with good rugged marine wood screws driven from the outside of the hull using at least two screws to the plank. Now remove the clamping bolts from the hull and sister frame. Plug the bolt holes with the proper size bungs set in epoxy glue. Plug the screw heads with bungs, cut flush, and sand. The bolt hole wood plugs should go all the way through to the outside of the sister frame where they may be trimmed off for a neat finish. If this method does not suit you, try laying up the sister frame using thin strips of wood and the same bolts through the hull and each layer of wood until a laminated sister frame has been built up thick enough to suit you. In this method you will have to wait until each layer's glue has set and cured before setting up the next layer.

The last method involves "spiling" off the curve of the hull inside and cutting a matching curve in new stock to make up a sister frame. Before you discard this method as too complex, you might consider the use of the "Weird Wonder" spiling gauge to take off the curve of the hull. (See p. 99, Figure 6.7.) This is nothing more than a section of cardboard with small diameter dowels pushed through the holes between the layers of outside paper. The homemade gauge is tacked up inside the boat close to the side of the hull whose curve is to be copied. Tack the gauge up vertically plumb using a bubble level. Push the dowels through the cardboard until they just touch the inside curve of the boat. The ends of the dowels now conform to the curve of the boat. Remove the gauge, being careful not to move the dowels, and again tack it down plumb to the stock to be cut and make a dot on the new stock at the end of each dowel. Connect the dots together with a continuous line and you have the curve of the hull which may be cut in the new stock with either a band saw or a sabre saw. This pre-sawn frame may now be bolted, screwed, and glued in place alongside the damaged frame.

Besides frames, planks, and stringers (braces that run down the length of the boat) there are the stems which are very vulnerable to all kinds of damage. In the area of the stern almost everything around the transom seems to come in for its share of trouble. In most cases these members, when damaged, will require complete removal and replacement. Not too long ago the only way most of these structually complex pieces of wood could be cut and shaped was by an expert with years of experience. With the advent of the marine epoxy glues however, it is now possible for the courageous do-it-yourself skipper, using laminating techniques and epoxy glue, to cut and lay up a new part no matter how complex. For example, consider the initial complexity of the "knee" brace (Figure 6.21). Traditionally,

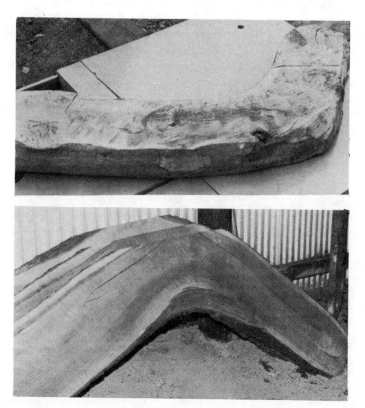

Fig. 6.21 Naturally grown, knees will be cut on band saw.

because of the extreme stress placed on braces of this type, they were cut from wood formed by natural growth of a tree either in the area where two branches forked or from the stump area where a root branched out from the stump. By using wood from these areas of the tree, the boat builder could cut and form a brace in which the grain of the wood followed the curve of the construction. Needless to say, it was not easy to find such a peculiarly shaped piece of lumber. The toughness of the wood and difficulties in initial cutting make lumber mills reluctant to cut them at all. When the do-it-yourself skipper needs to repair one of these complex forms his first problem is to obtain a reasonably accurate pattern of the original piece. Using laminating lay-up techniques and thin layers of wood (and plenty of time) he gradually builds up a new knee roughly in the shape desired. The final shaping can be done on a band saw or sabre saw, finishing with wood rasp and sanding discs. The finished piece is not only more handsome than the naturally formed wood, it is stronger by far and more resistant to rot due to the use of the epoxies. We have seen braces formed of alternate layers of different color woods that were works of art as well as highly functional structural members of the hull. These were finished in natural colors and varnished in clear spar varnish. No professional could afford to take the time to do this kind of work and make any money from repairing your boat.

To better illustrate this technique (Figure 6.22), we show a simple laying up form following the soft curve of the inner

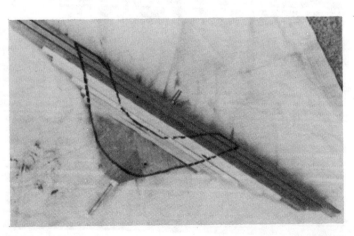

Fig. 6.22 Low cost easy method of laying up a knee brace which is much stronger than all others. Model shows approximate cut lines for band or sabre saw.

section of the knee. Several slices of thin, easily bent layers of wood were pre-cut on a table saw. Each layer was clamped into the form with epoxy glue and allowed to set and cure. When sufficient laminations were formed the knee was shaped using very little skill and a lot of patience. The advantages of the technique bear repeating. No particular skills in woodworking are needed. The resulting knee (or whatever structural member is needed) is stronger and more rot-resistant. It is costly only in terms of the time and patience required.

7
PLANNED PREVENTIVE AND CORRECTIVE MAINTENANCE AND THE FIBERGLASS HULL

Fiberglass boats have been around for some time and the techniques and methods, both in design and construction are constantly improving. Competition is fierce and is, for the most part, an advantage to boat owners. The buyer of a fiberglass boat is protected by the recent passage of the Boating Safety Act by Congress that charges the Coast Guard with regulating the quality of fiberglass boat manufacture insofar as it pertains to safety.

It may be convenient to think of the manufacture of a fiberglass boat as very similar to that of a fiberglass bathtub. The hull, when completely laid up of layers of various types of fiberglass cloth (or "roving"), is usually a complete integral unit with no seams to leak or to maintain. There is little wood to rot and few fasteners to corrode or rust, and with any kind of planned preventive maintenance the gel coat may not need paint other than the anti-fouling paint. Fiberglass boats are virtually impervious to either salt or fresh water, and the only damaging enemies are oxygen, sunlight, and man—not necessarily in that order.

Even so, some fiberglass boat owners complain that the fiberglass hull is hard riding and noisy in choppy water. We also hear complaints that the cabins of larger boats become damp from the condensation that forms on the inside of the hull as a result of moist air when the outside of the hull is in cool water. The dampness of all inside materials and fittings leaves a moldy odor, a problem that can be cured only by continuous ventilation and various drying systems. Other complaints are related to the cosmetics of the finishing gel coat, which the chemists have not yet perfected. Gel coats are more porous than one would think. They chalk up and turn dull in the sun and air and they are easily stained by rust from the wrong type of deck fasteners. A screwdriver or pair of pliers carelessly left on deck overnight will leave an indelible tracing of rust outline. So, the fiberglass boat is not totally free from preventive or corrective maintenance. For the most part, though, you can do much of the corrective maintenance yourself and *all* of the preventive maintenance.

MANUFACTURING METHODS

Fiberglass hulls are manufactured in a number of ways, none of them related to traditional wood boat building methods. The best of the fiberglass hulls are "layed" up by hand starting with a master mold that is the final shape of the finished boat (the biggest development cost item for fiberglass boat builders). A release agent (usually some form of wax) is sprayed into the mold to prevent it from sticking to the finished boat. Next the gel coat is sprayed onto the mold. The gel coat is a complex of various chemicals, dyes, and resins which give the boat its outside color and the beautifully finished coat that may be one of the heaviest contributors to the popularity of the fiberglass boat. The next several steps are similar except for the variation in the type of fiberglass cloth used in each step. Each boat maker has his methods of laying up a fiberglass boat. Glass fiber cloth comes as mat, open and fine weave, and heavy woven roving. In a high quality boat these are cut, fitted, and soaked in resin and a hardening catalyst (and sometimes other chemicals) layer by layer until the hull is thick enough. When it has cured (hardened), the hull is drawn out of the mold and work is begun on decks and interiors. During the lay-up, stiffening members of heavy wood, also covered with fiberglass cloth and resin, are set and bonded to the inside of the hull. These include such elements as engine mounts or beds, and the long stringers running from bow to stern on which cabins and flooring are mounted. In some construction the cabins and deck are molded in one piece like the hull and bonded together after each part has cured. The so-called sandwich type of construction is favored by several manufacturers. This involves placing layers of fiberglass and resin over cores of various materials such as plywood, balsa wood, and plastic foam. This method is said to greatly increase the overall strength of the hull but it can be a costly process.

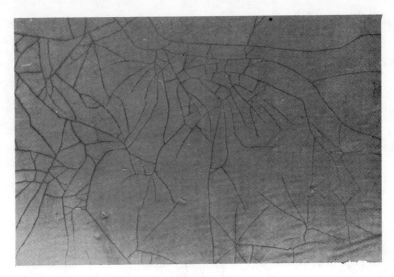

Fig. 7.1 Shown here "crazing" of gel coat.

During manufacture, there are a couple of things that occasionally happen to even the best of fiberglass boats and the result is corrective maintenance problems after the boat has been in service for a while. Fortunately, each of these problems indicates its presence by definite, observable clues. Among them are fine small cracks running in every direction called gel coat "crazing" (Figure 7.1). Another problem, similar to crazing, occurs when the gel coat wrinkles just like an area of a painted surface where too much paint has been layed on causing an "alligatoring" effect. Whenever there is too much resin at one spot during the layup of the hull, a condition known as "resin richness" exists. Under stress, a resin-rich area, without the reinforcement of the fiberglass, tends to develop small cracks in the gel coat (crazing). In an extreme case, the surface of the laminate might break down, become brittle, and cause a failure of the hull at that point. When the fiberglass is "starved" for resin, a void like a hollow forms in which there is neither glass nor resin. In this case, bubbles, lifting, and delamination of the glass layers may occur.

A large flat area in the hull might give you an "oilcan" effect—so called because of the clicking sound the bottom of an oil can makes as it is depressed to force oil out of the spout. The unsupported area will work in and out as the boat proceeds through the water. If attention is not given to this problem, at worst the laminate can fracture, or at least the gel coat can acquire a bad case of crazing. A simple effective cure for this problem will be described later in this chapter.

Voids in the laminate most often occur where sharp angles are formed in the hull, such as the point at the juncture of the trunk cabin and the deck or the inside of the hull where the side of the hull makes a near ninety-degree bend to become the bottom of the boat. Stress cracks frequently appear where deck fittings of metal that are placed under strain are bolted to the hull. All are repairable with the right tools, materials, and attitude.

GEL COAT MAINTENANCE

A substantial amount of your maintenance time on a fiberglass boat will be focused on both preventing and correcting problems with the gel coat. The gel coat, while by far the best thing that has happened for small craft cosmetics, *still* requires routine attention. There are a number of practical reasons to attend the gel coat, not the least of which is maintaining, even upgrading the boat's resale value. As mentioned in chapter 6, despite glossy appearance the gel coat is very porous. It stains easily, and in heavily polluted water the boat may acquire a mustache of discoloration at the bow near the water line. To quickly remove this mustache, use a light application of Naval Jelly which is normally used for rust removal on metal. Be sure the chemical has an acid base (to be safe, test it first on a very small area of the discoloration before tackling the whole job). It should attack the stain, not the gel coat itself. A number of commercial stain-removers work with varying degrees of success and require varying amounts of energy to apply and remove. Consult your dealer and read the instructions on the container *before* you buy. Another source of gel coat stains (and irritation) is metal objects carelessly left on the deck overnight. Rust from the metal gets into the pores of the gel coat and is a challenge to remove. Common lemon juice is rubbed onto the stain usually to remove the rust effectively. The gel coat is also very susceptible to oxidation and ultra violet rays of the sun. This causes a chalking, dulling effect of the shiny coat and can even cause a change in the color of the gel coat. A number of gel coat rubbing compounds containing a fine abrasive are available, and when

Fig. 7.2 Gel coat rubbing and waxing compounds.

applied as directed, with lots of energy, will restore the gel coat to its original gloss. Gel coat rub-down can be done using a power buffing pad. However, use extreme care not to rub a hole right through the gel coat which is only a few thousandths of an inch thick. When the gel coat has been rubbed down *do not delay* in getting a good coat of gel coat wax on right away. Use a wax that is formulated specifically for gel coats. Read and follow the directions for the results you want. Again, energy and patience are required. Manufacturers say that most of the complaints about product failure stem from the user's failure to read the directions — or if he read the directions, failure to follow them — or if he followed them, failure to do so correctly.

In spite of good care, gel coats suffer small problems that spoil their looks and open up the laminate to salt, air, and pollution from other chemicals. The restoration of the gel coat using the small gel coat repair kits is excellent. However there may be substantial problems in finding a gel coat repair kit that will *perfectly* match the color of your present gel coat. Even those kits made up by the company that manufactured the boat can only come close. The colors in the kit supplied will match the gel coat on your boat only if it is new. After a period of time the gel coat fades, chalks, and oxidizes, and the kit material does not match this effect — there is no way it can and there is nothing wrong with the colors of the kit. Try small samples of various kits until you get a satisfactory match — not perfect, but satisfactory!

Most gel coat repair kits come in two parts in tubes or small jars. The main ingredient is mixed with a smaller amount (usual-

Gel coat maintenance 127

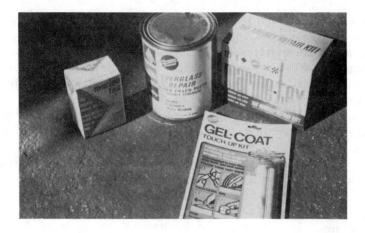

Fig. 7.3 Gel coat repair kits come in many colors and are good for small repairs to gel coat.

Fig. 7.4 To repair deep scratch, this small vee-shaped bit is perfect for routing out a vee-shaped cut in gel coat.

ly) of catalyst which causes it to begin hardening immediately. The directions on the container will tell you how much time you will have before the substance becomes unworkable. Often the ratio of catalyst to mix is critical, so do it right the first time. When the gel coat has suffered one or more of the more serious deep scratches which penetrate the gel coat completely, it may be necessary to use a vee-type routing tool, chucked into a small electric drill and cut a vee shaped trench all along the scratch into the laminate itself. This gives the repair materials more "tooth" to adhere to. Follow the mixing instructions and fill the drilled out crack with catalyzed gel coating. Fill the crack to a point just a little more than needed. Cover the repair with a clear plastic kitchen wrap and stick the edges down tightly with masking tape. When the repair material has set to a rubbery consistency (which you can test by touching lightly with a finger on the plastic covering) lift the edge of the plastic cover and with a single edge razor blade held at a low angle to the surface, trim the excess repair material flush to the surface. Replace the plastic cover and allow the repair to finish hardening.

When the repair gel has fully set, sand it and the area immediately surrounding with a wet and dry type paper which should be dipped in water from time to time. This will keep the gel coat material from clogging the abrasive and allow the cutting action to continue. Begin with a 220 grit paper, then use a 400 grit, and finish with a 600 grit. To finish, buff the area with rubbing compound and polish with a fiberglass gel coat.

When the area to be repaired is large (big as a quarter, and perhaps deeper), you can successfully follow the procedure described above. However, you should obtain some milled fibers along with the gel coat repair kit. Add the milled fibers to the mixed materials to make a putty. Fill the gouged area with putty slightly above the surface and proceed as before. In brief, the steps are: remove all loose material around the break with the edge of a putty knife; wash the area with an acetone soaked clean cloth; mix equal parts of milled fibers and matching gel coat; add catalyst to the putty according to instructions; apply the putty to the damaged area, carefully working it in to avoid air bubbles. Finally, buff, and wax or polish the patch. If the milled fibers in the patch do not match the color of the gel coat you will have to spray-finish the patched area with a gel coat paint.

PAINTING FIBERGLASS GEL COATS

A skipper who believes in PPM and practices it religiously will enjoy many seasons of satisfactory wear from the gel coat on his fiberglass boat—just how many seasons cannot be determined as all of the factors are not entirely within his control. Suffice it to say it will be a lot more than the boat skipper who does not believe in PPM.

If you have acquired a used boat whose gel coat has gone beyond the point where rubbing compound and elbow grease help, or if numerous scratches, gouges, and nicks make this

Fig. 7.5 New fiber glass paints are designed for fiber glass only.

procedure impractical, you should consider *painting* the boat. While in the past, some second thoughts *might* have been in order, due to the advent of new products and techniques, it is no longer difficult. There are a number of paints now available that have been compounded chemically for the sole purpose of painting a fiberglass boat. Many of these paints are themselves an epoxy-based or similar material. They are no more tricky to apply than paint for a wood hull. The most difficult part of painting a fiberglass boat is in the surface preparation, which is far more crucial than wood surface preparation. Fiberglass paint reacts to wax and grease and won't adhere to them. The smallest trace of old wax used as a releasing agent when the boat was first manufactured must be removed along with any of the subsequent wax polish coatings, and any grease and oil from polluted waters. Except for the drudgery involved, this is easily done by relentlessly scrubbing the entire area to be painted with a powerful detergent such as Oakite, followed by much rinsing. Further scrubbing with acetone should degrease the surface satisfactorily. The rest of what you need to know about painting fiberglass will be found in chapter IV, the instructions on the container, and your own experience.

HINTS ON WORKING WITH FIBERGLASS AND RESINS

As you might expect in working with anything that is as durable as fiberglass reinforced plastics, there are some precautions that must be taken if you decide, as we hope you will, to do much of your own maintenance on a fiberglass boat. Many people have a sensitivity to fiberglass, which causes their skin to itch. In any case, fiberglass dust particles should not be inhaled. So, when dry sanding fiberglass be sure to wear a dust mask. Many of the thinners, catalysts, and other chemicals may emit toxic fumes which can be a skin irritant. They also present an explosive hazard. For this reason, do not smoke while working with fiberglass materials. It is sensible to wear gloves for two *very good* reasons. First, there is a possibility that chemicals may splatter and irritate the skin (and they will not wash off easily). Second, the skin oil from your hands may ruin an otherwise perfect repair job!

Fiberglass cloth comes in a number of forms and each has its own application. Select the right material for the job. The very heavy weave of glass called roving provides great strength and is usually used inside the boat where its rough texture won't show. You will find roving layed up over stringers in the bilges for bracing and engine mounts. There are chopped strands of fiber that are used for both repairs and laying up the original hull. Milled or finely chopped fiber is available to use in making repair putty. An unwoven mat of short strands is available. The short strands

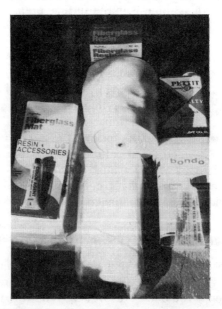

Fig. 7.6 Fiberglass cloth, mat, resin and hardener.

are laid in all directions and held together with a sizing that effectively soaks up a lot of resin for easy repairs. The random directions of the strands give the mat greater strength. Finally, there is finely woven regular cloth which is used as the layer next to the finishing gel coat. The primary thing to avoid when you are shopping for fiberglass cloth to do a repair job, is bargains. Some cloths were manufactured for other purposes and often have a sizing on them that will *prevent* the resins from sticking, while others are just not suited for the marine environment.

RESINS

We are using the word resins all-inclusively to include everything that isn't glass. Fiberglass chemistry is very complex and even a thorough knowledge of that chemistry will not help you one bit when it comes time to fix your boat. There are two terms that you must be familiar with—"shelf-life" and "pot-life." The chemicals used in some resins are very active and once they have been compounded and placed in the container, they must be purchased, mixed, and used within a certain limited period—that is, they have a limited shelf-life. Not all of these chemicals have this limitation and most containers include a statement and the last date on which it must be used. As for pot-life, this refers to the time that you have left in which to use the resin *after* you have mixed it with the catalyst. Once the catalyst has been added to the resin, the setting and curing process begins immediately, and the time that the mix is workable (soft putty-like) is limited. So, before you start to mix the resin and catalyst, make absolutely sure that everything is in readiness for this step. Once again, read the mixing directions. This advice applies to any of the marine glues and bonding agents including those that *are not* two-part types.

When you mix resins, mix up just one tablespoon more of catalyzed resin than you really need. If you can, work in small batches, mixing up and using the resin as you go along.

TOOLS FOR WORKING WITH FIBERGLASS

A good pair of scissors is essential for cutting fiberglass cloth. A rubber or plastic squeegee can be found in an automotive parts store. For a small job, use one of those rubber spatulas with the handle cut off.

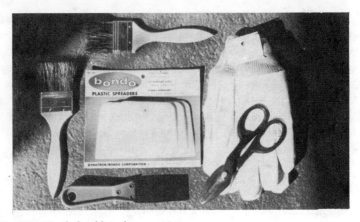

Fig. 7.7 Tools for fiberglass repairs.

An electric drill with a sanding, buffing and polishing pad attachment is almost a must in fiberglass repair. If it is any kind of a big job, consider renting a heavy duty buffing and sanding machine. If you do not already have an electric drill, you'll need one anyway. Buy the best you can afford but don't go overboard for one of the industrial types unless you plan on going into fiberglass repair as a business.

Some clean white cloth is needed for washing surfaces with acetone.

A sanding block with a soft rubber base will be most useful for smoothing out flush surfaces.

Two putty knives, one with a wide blade and the other a standard size are necessary. If you use them to spread catalyzed resin, clean them immediately after the spreading is finished.

Some kind of clear plastic sheets such as Saran wrap, plastic food bags and the like will be used with regular masking tape an inch or so wide when making repairs. You will also need a package of single edged razor blades to trim semi-hardened resins.

MATERIALS

Materials depend on the repair job to be done. However, you will be needing various sanding discs for the drill, various grits of sandpaper for the hand sanding finishing and a quantity of cleaner such as acetone or styrene. However if you do buy acetone, purchase no more than you need because it is very volatile and is explosive in vapor form. This warning is more applicable to the thinner/cleaners you will be using. A plastic bucket and clean water will be required and useful. Keep a CO_2-type fire extinguisher handy at all times. If you must work inside the boat with any of these materials, open the boat wide to the ventilating air and avoid breathing the fumes for long periods of time.

Depending on the repair, you will need small quantities of resin, catalyst, milled fibers, glass cloth, mat and roving. A quantity of the wet and dry type of abrasive papers in 100, 220, 400 and 600 grits will cover nearly all situations.

CORRECTIVE MAINTENANCE

Painting Fiberglass Boats

Normally, since a fiberglass boat hull is protected with the finish gel coat, painting over the gel coat must be considered a corrective action. When hairline checking, blistering, damage caused by the undercutting of the gel coat by water penetration through pin-holes, crazing caused by impact and stress, scratches and chipping have all brought the gel coat to the point where simple rubbing and waxing do not make it look any better, there are three alternatives: one, you can sell it as it is; two, you can paint it and sell it; or three, you can paint it and keep it. If you decide to paint it, don't decide to either sell or keep it until the paint job is finished and dry. You might be much surprised at the results (if done correctly). Although painting was covered in chapter IV we shall repeat the recommended procedures here to point out the small differences involved as opposed to the wood hull.

First, consider resurfacing the gel coat altogether when problems (listed previously) are present that are small enough not to warrant their actual repair with gel kits, but numerous enough to spoil the overall appearance of the boat. To bring the gel coat surface back to smoothness in preparation for the undercoat and final coats, the surface should be *glazed*. Glazing the surface with a polyester or epoxy compound especially formulated for this purpose is the best method of restoring the original smoothness. This compound is applied with a broad-bladed putty knife, allowed to cure and then sanded using wet type sandpaper. This allows the "feathering" of the hard compound into the surrounding gel coat. If the surface has only minor crazing or blemishes, the glazing may be omitted and the application of a proper primer such as a catalyzed epoxy undercoat will usually fill and fair-out minor surface imperfections.

In either case, the surface *must* be *most carefully* prepared. The main point is to completely clean off all dirt, wax, and grease. Scrubbing with a strong detergent and stiff brush is the best way. Rinse with clean water and allow to dry completely. The next step is to sand the surface with an 80 grit production paper to dull the surface of the gel coat. Wipe down with a clean rag soaked with paint thinner. The surface is now ready for the first coat of primer. Since the choice of the "system" (as the paints are now called) is up to you, your choice will dictate how it should be applied. One of the best and easiest for small craft boat owners is a silicon-alkyd type of fiberglass boat paint. The new polyurethane paints have been successful and should be looked into before you decide. However, shop around, seek advice from the professionals, and if possible try to find a boat that has been painted with the system you are thinking of using and ask the owner how he likes it.

REPAIRS

The following procedures for the repair of various traumas on fiberglass boats have been selected in much the same manner as those for wood hulls. Since we are unable to cover every possible needed repair, we have chosen those that will have near universal application with respect to method and techniques.

Through-Hull Puncture

This repair job assumes a complete fracture of the hull resulting from a puncture through the hull. The boat must be out of the water in order to effect the needed repair. First, inspect the inside of the hull to determine the extent of the damage to any structural member (if any) and to assure complete access to the damaged area from *within* the hull, because essentially, this repair is mostly done from the inside of the hull. A somewhat different technique will be required when the inside of the puncture is inaccessible due to internal structures. It would be helpful in arriving at the decision whether to work from the

Fig. 7.8 Three different types of fiberglass hull punctures.

inside or outside to read through these instructions first, then decide. Assuming that you will be able to work from the inside of the boat and there is no damage to any inner structural members, the next step is to assess the extent of the hull damage so that you may estimate your materials needs. Include in this estimate the fact that few if any punctures will be clean neat holes and

Fig. 7.9 Example of how cardboard backing plate is attached to hull with masking tape (note plastic covering over cardboard).

you will be cutting back several inches of damaged hull into sound laminate. Determine if the last layer of glass cloth showing inside the boat is heavy weave material. Use the disc sander (24 grit disc) to sand down the *outside* area around the fracture. The idea is to form a concave or cup-like area that you will fill with alternate layers of resin soaked fiber glass mat.

Assume for the balance of these procedures you will be working only from the inside of the boat and that the outside has *not* been sanded down, only the inside. The next step is to wash the entire cut out area thoroughly with acetone. From this point on do not touch any materials or the repair area with your bare hands — put on gloves. Now fabricate the *outside* backing plate, which is nothing more than a handy piece of cardboard cut to fit completely over the cut out area. Cover the side of the backing plate that will face the hull with plastic sheets. Stick it to the backing plate with masking tape. Apply the backing plate to cover the cut out area and stick it to the hull with masking tape. If the area to be repaired is curved, the cardboard will not do as it cannot smoothly conform to the curve of the hull. Instead, use a piece of thin aluminum plate. The best and cheapest source for this is one of those disposable large roasting pans found in many hardward stores. Cut out the bottom flat section, cover with plastic wrap, smooth to the shape of the hull and stick down with masking tape. Following the instructions on the container for mixing, mix a quantity of gel coat and catalyst on a sheet of cardboard. Working from the inside of the boat, put a thin layer of gel coat catalyzed putty on the backing plate using the rubber squeegee. Make sure there are no air bubbles in the putty after it has been applied and be sure the putty overlaps or fills the cut out area to the edge of the hull and old gel coat. The gel coat should "gel" before the repair commenses. Cut a piece of fiberglass mat to fit in the cut out hole. Place this piece of mat on a sheet of cardboard and using a paint brush *saturate* the mat with catalyzed resin. Work out any air bubbles with the brush. The whole trick in working with fiberglass is to thoroughly saturate it with resin and then get out all the air bubbles. It must not be resin rich or resin starved. When the first piece of mat is ready, place it over the puttied gel coat and tuck it gently into place with the brush. The first layer will help the gel coat to cure away from the air, and still be a little "tacky" so that the layer will stick. Cut, saturate, and debubble each succeeding layer of mat and tuck into place, building up the patch. If the patch is more than three-sixteenths of an inch thick (to match the hull thickness), lay up just half of the mat layers and let them harden for an hour or so until the surface feels leathery and forms a good base for the rest of the layers. Mix a new batch of resin and catalyst, cut, saturate, and debubble the remaining layers until the patch is flush with the hull. The last two or three layers laid up after the patch is flush should overlap by an inch or so the cut out area of each previous layer. The last layer, inside the hull, should be woven roving, well saturated and cut to fit the sanded area inside the hull. Leave it to set and cure completely, resisting the temptation to poke at it from time to time. Spend the time cleaning the tools and picking up. When the curing and hardening is complete, carefully remove the backing plate from the outside of the boat. If there are any pits or roving, purchase a section large enough to cover the cut back puncture with an overlap of three to five inches. Fiberglass mat is recommended for the layers of the repair itself since it will soak up resin easily and can be squeegeed to a smooth finish. You will also need a piece of smooth and fine-woven fiberglass cloth which, although it will not be a part of the final repair, will be used during the repair. You will need resin, catalyst, gel coat, abrasive papers, thinners, and acetone. Shop for these materials by reading the directions on the labels *before* you buy to determine if the container amount will do the job. The materials are expensive but it is best to miscalculate on the side of generosity. When you have assembled all the *materials* you will need, including those you *might* need, gather up the *tools*, again, including those that you *might* need.

Now mark out the area around the fracture with a felt tip pen or crayon on the outside of the boat to trace a cutting line around the fracture, cutting back *only* to sound hull. Note that if the fracture is a long narrow one, it will not be necessary to cut back with the sabre saw. Since a hole through the hull already exists, it will not be necessary to drill a pilot hole for the saw blade. Take a deep breath and start cutting. When the damaged hull material has been cut away, the next step is to sand down the surface around the inside of the cut out area back to two or three inches and deep enough that one layer of final woven roving cloth will just fit flush to the old surface. Again, if the puncture was long and narrow and it was not necessary to cut out damaged hull, other imperfections in the gel coat patch can be touched up with gel coat repair kit. When the gel coat patch is fully set and cured the final and finishing steps require only patience and some drudgery if you really desire a high gloss finish over the patch. With the sanding block of a 100 grit wet and dry sandpaper, used wet, sand the gel coat smooth and flush with the rest of the hull. Don't be in a rush. Wash the sanding dust off frequently to see how the surface is coming. Keep going up in grit numbers (finer and finer size) until you finish off with a 600 grit paper. Finally, rub the patch down with a soft cloth and fiberglass rubbing compound to bring the surface to a glittering shine. Wax and buff and the repair is complete.

In those cases where there is an obstruction inside the boat and you are unable to work from the inside, the technique called for is a "blind patch." Essentially, this is the same as the preceeding method except the backing plate must go on the inside and you work from the outside of the boat. The first step is still the cutting away of all damaged laminate, sanding back from the edge of the patch two or so inches on the outside with the disc sander. Obviously, you will not be able to use the disc sander on the inside to sand back from the inside of the hole to be patched. If the hole is too small to get your hand into, then you must use the milled (chopped) fibers mixed with catalyzed resin to form a putty. If the hole *is* large enough to get your hand through, sand down the edge of the hole inside to about two inches around the perimeter of the hole. As before, wash both the inside and outside of the sanded area with a clean cloth soaked in acetone and don't touch any part of the area with bare hands after it has been washed. Cut a piece of cardboard to fit *inside* the hole at least as large as the sanded area. Poke a length of stiff wire (copper if you have it) through the cardboard from the outside and back again so that it forms a "U" in the cardboard and can be used to pull and hold the backing plate against the inside of the hole. You may have to take some time to get it to fit tight and flush against the inside of the hole, but be patient and do it right. When you are satisfied that the backing plate does indeed fit tight and flush against the inside of the hole—remove it! Pull the wire out of the plate. Cut a piece of fiberglass cloth *and* mat the same size as the cardboard. Put the cloth on top of the cardboard and the mat on the top of the cloth. Wet the cloth and mat thoroughly with catalyzed resin, taking pains to get the bubbles out. Reinsert the wire into the cardboard, cloth, and mat. Put the cardboard back in the hole and pull the wire forward until the patch is pulled flush against the inside shoulders of the hole. It is essential that the cloth and mat patch inside be in contact all around the edge of the hull. Put a stick or board on the outside of the hole and twist the wires around the stick to hold the patch in place while it is setting and curing. When the patch has hardened and cured, separate the wire from the stick by using a pair of diagonal pliers (wire cutters with sharp points) to cut the wire as close to the patch as you can. With the disc sander, sand off the wire that shows. If the balance of the hole to be patched is too small for layers of mat or cloth, the putty method is both quick and easy. With this technique you mix up a putty of milled fibers and catalyzed epoxy resin and apply it to the hole to be filled until it just barely fills the hole leaving only enough room for a thin layer of gel coat. Allow the putty to set and harden. Mix gel coat putty with catalyst to form a thick creamy substance and spread evenly on the outside of the hole with the rubber squeegee. Cover with a backing plate of cardboard and plastic wrap and seal with masking tape. When the gel coat has hardened, finish as in the previous procedure.

    This procedure is also effective for filling in a previously cut hole that is no longer needed (such as a through-hull fitting that has been moved or discarded). Keep in mind that repair to a fiberglass hull should not be delayed just because it is fiberglass. If water gets into the laminations there will be trouble later on. Get the boat out of the water and get to it right away. Once you get the hang of working with the various forms of fiberglass and resins the chances are you'll find it easier than working on wood hulls.

CURING AN OILCANNING PROBLEM

An area of the hull that is oilcanning can be repaired relatively easily. The first step is to positively identify the area of the hull that is giving the effect by watching it as the boat is underway in the water. Mark the area with crayon or chalk. Naturally, the area must be accessible and not under fuel or water tanks. Find a cardboard mailing tube or paper towel tube. Using a razor sharp mat knife, cut the tube in half length wise. The length of the tube should be the length of the oil canning area or as close to it as possible. If the area causing the problem is curved you will have to notch the cardboard tube along its length so that it will bend and conform to the shape of the hull. Using the power disc sander and a 24 grit disc, sand the area of the oil canning effect to four or five inches either side of the cardboard tube form. Cut, saturate with catalyzed resin, and lay up several layers of fiber glass mat, cloth, and roving over the cardboard form. Each layer should be an inch or so larger than the preceding layer so that the edge of the brace tapers nicely into the hull. Be sure that the ends of the tube form are covered and sealed. When the brace has cured and hardened you'll find that the oil canning has disappeared.

The same procedure can be applied in laying up a deck brace. It is rare indeed that fiberglass boat builders feel the need to brace decks and cabin roof tops. For this problem, begin by fabricating a wood deck beam in the form of a "T." This can be made of plywood or any good quality strong wood. Since it is going to be covered completely with fiberglass cloth you will not have to dope it with wood preservative. Lay angled fillets of wood along the angle formed by the joining of the two pieces of wood that form the "T." Sharp angles form voids in fiberglass and should be studiously avoided. The wood "T" can be set up with glue and wood screws and attached by any convenient means to the overhead where the deck or roof is bending. The area on either side of the "T" brace or beam is sanded back three or four inches either side of the beam to give the fiberglass cloth a good bonding area. After washing the sanded area and the wood carefully with acetone complete the support beam by laying up several layers of mat and cloth using the usual techniques and precautions. When all the fiberglass has set and cured it may be sanded smooth and painted to match the rest of the overhead. This procedure is effective for bracing anywhere on the boat. As you become more proficient in the use of fiberglass materials

Curing an oilcanning problem 141

Fig. 7.10 Model of cardboard mailing tube cut to form brace to prevent "oilcanning" of hull section.

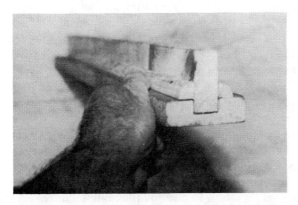

Fig. 7.11 Model of wooden deck brace before installation and the laying up of glass cloth/resin. Voids are prevented by using ¼ rounds to fill 90° angles.

and resins you may tackle some sohpisticated boat improvements such as a hand layed up bait, ice, or fish tank for stern area, rope and chain lockers and so on.

8
PLANNED PREVENTIVE MAINTENANCE AND CORRECTIVE MAINTENANCE FOR ALUMINUM HULLS

In spite of an early and abortive attempt to use aluminum as a boat building material, the idea persisted and the industry soon developed an alloy of magnesium and aluminum which can stand the gaff of life at sea. Aluminum, strong and light in its pure state, becomes yet stronger when alloyed with other metals. (The alloy for boats is the 5,000 series, the various digits indicating the particular alloy mixture used.) One unique advantage of aluminum is that the surface, exposed to air, forms a tough, clear, and non-porous coating of aluminum oxide. Thus, no other protective coating is required on the aluminum hull. Painting an aluminum hull is only for cosmetic reasons—not for protection. As a hull material, it is very light-weight in relationship to its strength. The world's fastest racing yachts have aluminum hulls, masts, and spars. It is resistant to dents and because of it's ductibility, will stretch before it breaks. Aluminum will not cause sparks when struck and is not a fire hazard. It is impervious to rot and ship worms. Since it is a non-magnetic metal it does not affect a compass and cause deviation problems. Construction costs are lower, and because of its ductibility it is easily worked. It can be bent, shaped, rolled, sheared, and stretched using tools less complex and less expensive than those required for steel and wood. For fastening, special welding techniques have been developed, along with special rivets.

The aluminum hull is no more perfect than the fiberglass hull. Much like fiberglass it is very noisy when under way. Aluminum is a very "active" metal and one must be constantly on guard against galvanic corrosion. Although it is resistant to acid, the alkaloids create with it. Skippers who trail an aluminum hull boat over the road must guard against the effect of the special salt-like chemicals spread on the road to melt ice and snow. In addition, the aluminum hull must be stored well clear of the bare ground to avoid corrosion. Aluminum is second only to copper as an electrical conductor, and as a result, it is also susceptible to galvanic action. An aluminum hull boat tied up near a steel hull boat might end up as a sacrificial anode to the steel hull if all equipment aboard is not carefully polarized and isolated from strict *electrical ground* by an isolation transformer. All metal attachments to the hull (such as drive shafts, struts, water intake strainers, valves, and propellers) must be insulated from direct metal-to-metal contact using "pads" or washers of plastic. More careful attention to the maintenance of sacrificial zincs is a *must* on aluminum hulls and the owner/skipper needs to be very knowledgeable on the subject of galvanic action, its cause and cures. Finally, in spite of the many advantages of aluminum, marine organisms will quickly attach themselves to the aluminum hull, and such a hull, if operated in salt water, must be protected with bottom paint—not just any bottom paint, but one that *does not* contain copper as the active ingredient. Copper based anti-fouling paint, painted directly on the aluminum surface of the hull, will start an immediate and raging case of galvanic action corrosion. Balancing the advantages and disadvantages, it may well be that the aluminum hull is the closest to being "maintenance free."

With respect to PPM there are but two concerns: first and foremost, the control of galvanic action, already touched upon; and second, the critical application of anti-fouling paint protection. In a very important way, these, as far as aluminum hull boats are concerned, are interrelated.

ANTI-FOULING PROTECTION

Briefly reviewing the factors of the problem, marine organisms, whether plants or animals, attach themselves to the hull of a boat as a place to live and not to chew on the boat as a ship worm might. As a result, they have no predjudices with respect to the hull material—wood, fiberglass, aluminum, and steel—are all equally attractive. Thus, a boat that is going to be kept in salt water, even brackish water, must be afforded some protection **and** the prevention against growth or attachment of marine organisms to the hull. The ingredient in bottom paint that provides this protection is a metal in a powdered form suspended in the paint. Applied to the hull it reacts to salt water or it leaches to the surface of the paint and poisons the growth since it is highly toxic. Previously the effective metal used was mercury, which has since been replaced by the environmentalists because, as they claim, mercury never changes its nature. Once in the water it can be ingested by small marine organisms which in turn would be eaten by larger organisms until the mercury ends up in food fish to be ingested by humans. so mercury is out, as well it should be. By far the most popular toxic agent now in use is copper followed distantly by tin. Copper based paint *cannot* be directly applied to the bottom of an aluminum hull without disastrous results. The copper in the paint, the sea water, and the aluminum hull involve themselves in a raging case of galvanic action with aluminum ending up looking more like a lace curtain than a solid sheet of metal.

There are three solutions to the problem. The first and by far the simplest is to place the boat in the water *only* when it is going to be used and immediately remove it from the water during non-use. When this is not feasible, a "system" of bottom painting must be used. In the simplest terms, this involves several coats of paint carefully laid up in such a way so as to form a "barrier" coat of neutral paint interposed between the aluminum hull and the copper bottom paint. This system will work fine as long as there are *no scratches* in the paint that penetrate the layers of paint down to the bare hull. Should this happen, a pocket of galvanic corrosion can begin resulting in pin hole penetration of the hull. The skipper with the aluminum hull and a copper bottom paint system must pay close preventive maintenance attention to the anti-fouling paint on his boat. The best solution to date seems to lie in the application of one of the new anti-fouling paints in which the toxicant metal is "organo-tin." Galvanic action between this paint and aluminum is virtually nil. A new hull must be degreased and coated with three coats of an epoxy primer paint before the bottom paint coats go on.

Not long ago, one of the big rubber companies announced a new anti-fouling product consisting of thin rubber sheets that were to be bonded to the hull. The toxicants are a part of the rubber but would be insulated from the hull. Since the sheets can only be installed by authorized dealers, its cost and low availability restrict its use to "yachtsmen." Finally, during the fuel crisis, the Navy scientists came up with a bottom paint which is said to last for years since the toxic material is passive until a marine organism tries to attach itself to the hull.

GALVANIC ACTION

The second part of the problem, galvanic action alone, does not logically apply to the smaller aluminum hulls such as canoes, skiffs, and flat bottomed duck boats. These craft are powered by outboard engines that are mostly assembled of aluminum also. It is the larger hull in the cruiser class that has an inboard engine, bronze shafts, propellers, and struts, and other through-hull fittings of various metals that faces the ever present danger of galvanic corrosion. One (or both) of two things must be done. Either the fittings which are attached to the hull *must* be insulated electrically by installing heavy pads of plastic between the fitting and the hull, or several sacrificial zinc anodes must be bonded tightly to the fittings. These anodes must be carefully watched and renewed faithfully as they are consumed. This may not be the easiest thing to do when the boat is in the water.

The skipper with an aluminum hull houseboat often installs a dockside electrical power system to support all the on-board appliances. If the greatest care to maintain the correct connecting procedures is not taken, he may find that his hull becomes a great big sacrificial anode to some other boat or a nearby mass of steel.

ALUMINUM HULL CORRECTIVE MAINTENANCE

Tools. You will need a large rubber mallet, a small plastic-faced mallet, and one or two backing irons.

Simple dents. Sometimes, but not often, simple dents can be removed by no more complex a procedure than pressing them out from the inside with a determined thumb. When they are too stubborn to respond to this treatment, use a rubber mallet from an automotive parts store. (With an aluminum hull, you'd be wise to have one anyway.) As you apply the rubber mallet on the inside of the hull, assuming that the dent is accessible to the mallet, you will require someone to hold a "backing iron" to the outside of the hull over the dented area. A backing iron is nothing more than a curved piece of steel which serves as an anvil on which the dented hull is pounded back to shape. Both of these tools can often be rented from the tool rental shops. These same shops may have a dent pulling tool. Use of this tool involves drilling a hole through the hull at the center of the dent and attaching the tool through the hole. A heavy weight is slid smartly back to the end of the tool much like hammering. This has the effect of *pulling* the dent out.

Keep in mind that the malleability of aluminum decreases as it is worked—that is, becomes work hardened. The aluminum tends to stretch when you attempt to straighten out a dent, resulting in too much metal at the end, or a wrinkle. To avoid this problem, drill a modest hole at the apex of the dent, thus giving the metal a small area to stretch into as it is hammered back into shape. Then, after the dent has been straightened, fill the hole by riveting two thin aluminum washers, one inside and one outside, using a "pop" riveting gun (illustrated on page 102).

SPECIAL PROPERTIES OF ALUMINUM HULLS

Another factor must be kept in mind when repairing and painting with modern sealants, resins, and bonding agents. The rapid formation of an oxide on the surface of aluminum, while providing its own protection, also prevents the adhesion (sticking) of many other valuable protective substances. When you are considering the use of a sealer or a filling putty for finishing off the repair of a dent or tear in the aluminum hull, select a product whose label or instructions clearly states that it is a satisfactory adherent to aluminum. Zinc chromate putties, silicon based sealants and polysulphide caulking compounds in tubes are known to be effective.

Paint removal must be done with care. For example, sand blasting is quick and effective but must be done with clean sand not used on ANY other metal. Otherwise there is the risk of galvanic action from the other metal sticking to grains of sand imbedded in the aluminum. If a paint remover chemical is to be used, select one that does not attack aluminum.

Welding is now considered to be the best method for fastening aluminum. It does however, require rather sophisticated

Special properties of aluminum hulls 147

equipment and a good degree of skill and welding knowledge. For this reason leave welding to the pros.

Tears

Because of its ductibility aluminum does not easily tear. It *has* been known to happen however. To repair a tear, first "ding" the edges of the tear flat and flush with the rest of the hull using a soft lead-faced hammer. "Dinging" is a method used to straighten dents in auto bodies. Someone will have to hold a heavy block of wood against the outside of the hull while you tap slowly and carefully inside the hull. When the puncture is satisfactorily flush, the ragged edges must be sanded off with a garnet paper sanding disc. Drill small "stopping" holes at the ends of the tears to prevent the spread of the tear into good metal. Next, bond a sheet of tracing paper to the outside of the hull over the tear using masking tape and outline an oval-shaped patch on the paper with a felt tipped pen. Cut out the paper template and stick it to a piece of thin aluminum plate stock of the same thickness as the hull of the boat. (The aluminum stock for the patch can be located by using the yellow pages of the phone book and checking with local dealers until you find one who stocks marine type sheet aluminum.) Clamp the sheet stock to a scrap of plywood and transfer the paper template outline to the stock, again with a felt tipped pen. Chuck a special fine toothed blade into a sabre saw and cut a backing patch from the sheet stock. Temporarily affix the patch to the hull by drilling (and deburring) holes at each end of the patch and securing with nuts and bolts to hold the patch in place while applying the rest of the fasteners. Remove the nuts and bolts, give the patch a layer of silicon sealing compound around the edge, and reset in place using the nuts and bolts. Rivet the patch by drilling three holes at a times and installing pop rivets after the holes are deburred. In order to prevent any stretching or wrinkling of the patch, the first three holes should be made in the upper right section of the patch and the rivets installed. The next series of three holes spaced close together should be drilled in the lower left section of the patch. In this manner the perimeter of the patch can be drilled, deburred, and pop riveted. Excess sealant should be wiped off using the solvent that came with the tube. Apply zinc chromate putty to the edges of the patch and sand with a fine grade of garnet paper. After repainting, the entire repair will be barely discernible and the hull will be fully watertight.

Tightening Seams

Aluminum hulls are manufactured with aircraft type rivets holding the seams together. While they rarely tear out or loosen, this has been known to happen. If you are lucky, a loosened seam can be retightened by using an upsetting tool (similar to a backing iron) and tightening the rivets with a hammer. When the rivets have been destroyed by de-formation, they will have to be removed by drilling or driving them out with a punch. Replace them with pop rivets and a daub of sealing compound in the seam. Internally, much like fiberglass boats, there are few problems with ribs, stringers, and other braces. Fortunately, many hardware stores now carry various forms of aluminum stock in the shape of rod, angle stock, threaded rod, and so on. While not specifically designed for marine use, they all can be used inside the boat for braces and repairs to fixtures keeping in mind the problems of galvanic action and corrosion. In addition, and if it is out of sight, there is no reason not to use wood to effect repairs inside the boat as there is likely to be a lot of it on the aluminum hull anyway.

Fig. 8.2 Two views showing aluminum hulls with aircraft type rivets.

Index

Adhesives (glue), 105
Anti-fouling paint, 35, 143

Bilges, 92
   garboards, 4
   painting, 80
   through-hull fittings, 92
Boat box (contents)
   belt, 48
   battery, 48
   cotter pins, 49
   distributor spares, 49
   dry cells (batteries), 50
   fasteners, 49
   first aid kit, 50
   flash light, 49
   gasket paper, 49
   hose, 48
   lamps, fuses, wire, 49
   nylon twine, 50
   RTV, 49
   sail needles, palm, thread, wax, 49
   sandpaper, 49
   sparkplugs, 48
   stove fuel, 50
   wiper blade, 50

Checklists, 39

Decking, 93
Defenses (against the enemies), 32
   air, 59
   galvanic action, 60
   man, 62
   plants, animals, 62
   rot, 61
   rust, 61
   salt, 60
   sunlight, 57

Enemies (description)
   air (corrosion), 28
   galvanic action, 29
   man, 29
   plants, animals, 31
   rot, 30
   rust (salt), 28
   sunlight, 27

Fastenings, 87

Index

Galvanic action, 29, 60, 145
Garboards, 4, 35

Hulls (aluminum), 142
Hulls (fiberglass), 123
Hulls (types), 24, 33
   displacement, 24
   modified deep vee, 26
   planning, 25
   vee, 25
Hulls (wood construction), 85
   carvel, 85
   lapstraked, 85
   plywood, 86

Job analysis, 41

Knees (repair), 119

Materials (for PPM), 55
   fiberglass, 56
   glue and bonding agents, 56
   lubricants, 55
   paint, varnish, wax, 55
   preservatives (wood), 57
   wood, 55

Painting, 63
   aluminum hulls, 78
   aluminum masts, spars, 79
   bilges, 80
   brightwork (varnish), 80
   brushing techniques, 76
   cabin interiors, 80
   decks, 80
   estimating needs, 67
   fiberglass, 128, 133
   final coats, 76
   hull interiors, 80
   new wood, 76
   outdrives, 79
   planning, 63
   previous coats, 77
   removal methods, 71
Painting tools, 68
   brushes, 72
   buckets, 73
   putty knives, 70
   rollers, 74
   scrapers, 68
   sandpaper, 68
   tack cloth, 71
   wire brushes, 68
Painting transoms, 80
Planking, 90
   checking, 91
   seams, sealing, 90
   wood, 91
Planned preventive maintenance (PPM), 1, 23, 32

Repairs to aluminum hulls, 143
   dents, 145
   painting (anti-fouling), 143
   seams, 149
   tears, 147
Repairs to fiberglass hulls, 122
   gel coats, 125, 128, 133
   materials, 129, 132
   oilcanning, 140
   punctures, 134
   tools, 131
Repairs to wood hulls, 107
   carvel planks, 112
   frames, 116
   knees, 119
   lapstrakes, 109
   plywood, 107

Scarph joint, 4
Shafts, 5
Soft spots, 3
Stem, 34
Stern, 4

Systems (that make up a boat), 3
    electrical, 9
    hull, 3
    life support (comfort-convenience), 13
    propulsion, 5
    safety, 15
    steering, 13
    support, 21

Tools, 44
    boat box, 45
    hand, 51
    power, 52

Wood rot, 4